# Landlord and tenant

Hundreds of millions of tenants live in Third World cities; in many cities they constitute the majority of households. Despite their numerical importance, we remain ignorant of who this large mass of people is and of the conditions in which they live. We are even more ignorant about those who provide rental accommodation, even if we have inherited strong prejudices about the characteristics of the typical landlord. This book attempts to summarise what we know about rental housing in Latin America and makes comparisons with the situation in other areas of the developing world.

It uses survey data and in-depth interviews to compile a detailed picture of landlords and of tenants in Mexico. The authors explore why landlords invest, and sometimes fail to invest, in rental housing. They also examine why tenants live in rental housing and under what circumstances they move into peripheral self-help ownership. They look at the relations between landlord and tenants and at how the state intervenes in that relationship. *Landlord and Tenant* also reviews Mexican housing policy, with its clear bias towards increasing home ownership, and explores ways of improving the quality and increasing the stock of rental accommodation.

This is the first book to be written in English about rental housing in Latin America, or indeed about rental housing in any Third World country.

**Alan Gilbert** is Professor of Geography at University College and the Institute of Latin American Studies, London. He has published several books and numerous articles on housing and urban development in Latin America. He has also acted as a consultant to several governments and United Nations organisations. **Ann Varley** is a lecturer in the Department of Geography, University College London. She has been working on housing in Mexico since 1980, with a special interest in government policies for improving the quality of self-help housing.

# Landlord and Tenant

## Housing the Poor in Urban Mexico

Alan Gilbert and Ann Varley

London and New York

First published 1991
by Routledge
2 Park Square, Milton Park, Abingdon, Oxon, OX14 4RN

Simultaneously published in the USA and Canada
by Routledge
270 Madison Ave, New York NY 10016

Transferred to Digital Printing 2008

Typeset by Laserscript Ltd, Mitcham, Surrey

*British Library Cataloguing in Publication Data*
Gilbert, Alan, *1944 Oct. 1* –
   Landlord and tenant : housing the poor in urban Mexico.
   1. Mexico. Housing
   I. Title II. Varley, Ann, *1958* –
   363.50972
   ISBN 0-415-05593-8

*Library of Congress Cataloging in Publication Data*
Gilbert, Alan, 1944–
   Landlord and tenant : housing the poor in urban Mexico /
   Alan Gilbert and Ann Varley.
     p. cm.
   Includes bibliographical references.
   ISBN 0-415-05593-8
   1. Rental housing – Mexico. 2. Poor – Housing – Mexico
   3. Housing policy – Mexico. I. Varley, Ann, 1958–  . II. Title.
   HD7288.85.M6G55   1991
   363.5'96942'0972 – dc20          90-35002
                                    CIP

**Publisher's Note**
The publisher has gone to great lengths to ensure the quality of this reprint
but points out that some imperfections in the original may be apparent.

# Contents

# Contents

# List of figures

# List of tables

# Glossary of Mexican acronyms

Mexico is a country full of acronyms. Hopefully, this glossary will help the reader to follow the text.

| | |
|---|---|
| AMPI | Mexican Association of Property Managers |
| BANOBRAS | National Bank for Public Works and Services |
| BNHUOPSA | National Bank for Urban Mortgages and Public Works |
| BNOPSA | National Mortgage Bank for Public Works in Urban Areas |
| CEPROFIS | Certificates for fiscal promotion |
| CNC | National Peasant Confederation |
| CNIC | National Confederation of Tenants and Settlers |
| CONAMUP | National Coordinator of the Urban Popular Movement |
| COPEVI | Operational Centre for Settlement and Housing |
| CORETT | Commission for the Regularisation of Land Tenure |
| CTM | Mexican Workers' Confederation |
| DDF | Department of the Federal District |
| DGHP | General Directorate for Popular Housing (Federal District) |
| DGPC | General Directorate for Civil Pensions |
| DPE | Directorate for State Pensions (State of Jalisco) |
| DPM | Directorate of Military Pensions |
| FOGA | Fund for the Guarantee and Support for Housing Credit |
| FONHAPO | Trust Fund for Popular Housing |
| FOVI | Fund for Banking Operations and Discounts to Housing |
| FOVIMI | Housing Fund for the Military |
| FOVISSSTE | Housing Fund for State Workers |
| IMSS | Mexican Institute for Social Security |
| INDECO | National Institute for Community Development and Housing |
| INEGI | National Institute for Statistics and Geographical Information |
| INFONAVIT | National Institute of the Fund for Workers' Housing |

| | |
|---|---|
| INVI | National Institute for Housing |
| ISSSTE | Institute for Social Security and Services for State Workers |
| PEMEX | Mexican Petroleum Company |
| PRI | Institutional Revolutionary Party |
| PST | Socialist Workers' Party |
| PSUM | Mexican United Socialist Party |
| RHP | Popular Housing Renovation |
| SAHOP | Secretariat for Human Settlements and Public Works |
| SEDUE | Secretariat for Urban Development and Ecology |
| SHCP | Secretariat for Finance and Public Credit |
| SPP | Secretariat for the Budget and Planning |
| UDI | Union for the Defence of Tenants |
| UID | Democratic Tenants' Union (Puebla) |
| UII | Independent Tenants' Union (Guadalajara) |
| VIS-R | Social-Interest Housing for Rent |
| VITEP | Housing for the Workers of the State of Puebla |

# Preface

This book forms part of a wider programme of research into rental housing which has been under way at University College London since 1977. The initial work formed part of a large project, funded by the Overseas Development Administration (ODA), focusing on public intervention, housing and land use in the cities of Bogotá, Mexico City and Valencia. An important objective of that research was to understand better the process through which tenants become, or failed to become, home-owners. Interviews conducted with some 210 tenants in Bogotá and Mexico City provided a means of comparing the socio-economic characteristics of renters with those of owners (Gilbert, 1983; Gilbert and Ward, 1985). It also provided some of the first data collected in Latin America on the identity of the owners of peripheral rental property. While this work was in progress, Michael Edwards began his doctoral study of renting in Bucaramanga, a middle-sized Colombian city (Edwards, 1981; 1982). The findings of these two studies provided the main conceptual framework on which the present study was based. Again funded by the ODA, the current research has sought further information on the tenant-owner transition but has also tried to investigate more profoundly the process of rental-housing investment, the nature of landlord-tenant relations, and the form of state intervention in the rental housing market. In parallel with this study, Gill Green carried out her doctoral research on landlords and tenants in Santa Cruz, Bolivia (Green, 1988a; 1988b).[1]

The point of this brief history is to underline the wider comparative basis into which this project fits. While the current book is based wholly on research in two Mexican cities, the broader objective of the programme is to provide the empirical base on which an informed theoretical understanding of the housing process in Latin American cities and the role played by rental and shared accommodation can be formulated. This work continues through a major grant from the Canadian International Development Research Centre which is funding a comparative project in the cities of

Caracas, Mexico City and Santiago de Chile.[2] This work is being conducted by local professionals from each city.

While the wider context is of major importance in our research design, this book is centrally concerned with the housing situation in Mexico. In itself, this is hardly an insignificant concern. Mexico, after all, contains several of Latin America's largest cities and what is arguably the world's most populous metropolitan area. With the urban population likely to have been growing at around 4.5 per cent during the 1980s, it is possible that Mexico's cities now contain 69 million people. Since somewhere between one-third and two-fifths of this number are renting or sharing accommodation, the subject of this book is of great importance to Mexico. Indeed, the Mexican government has recently recognised that rental accommodation is an option which it ought to encourage in its housing policies.

However, the choice of Mexico was not determined only by these considerations. Both authors already had strong links with the country and wanted to consolidate and widen their existing research experience. Ann had lived in Mexico City while carrying out her doctoral research on illegal housing development and tenure legalisation, and wished to extend her experience to other cities of the country. Alan had already participated in research in Mexico City and wished to widen his housing experience, previously focused on Colombia and Venezuela.

In carrying out the research, we have received a great deal of help from a multitude of individuals and institutions. The Overseas Development Administration funded us without any interference in our research strategy and method. The Mexican Ministry of Urban development (SEDUE) was helpful both in granting us permission to carry out the research and in sparing the time to explain different aspects of their housing policy. In Guadalajara, the Colegio de Jalisco and the University of Guadalajara's Faculty of Geography were especially helpful in providing Ann with local contacts and logistical support. In Puebla, the Colegio de Puebla and the Universidad de las Américas provided similar help. In London, University College and the Institute of Latin American Studies provided us with a wide range of financial, moral and institutional backing. In San Diego, the University of California provided Alan with a fellowship which gave six valuable months for contemplation on the issue of rental housing.

Numerous individuals should also be mentioned for their various kinds of help. In Mexico, Mario Carrillo, René Coulomb, Patrice Mele, Guillermo de la Peña, Daniel Vázquez and Andrés Zeromski all helped us immensely at different stages of the study. The contribution of the young Mexicans who helped Ann to carry out the questionnaire survey in Guadalajara and Puebla was considerable. Above all, we need to thank those many Mexicans who talked to us about housing issues or who patiently answered our questionnaire. Without all these people, we would not have a book today.

# 1 Introduction

Where do the poor live in Latin American cities? To many readers, the answer may seem to be obvious: poor households live in self-help settlements, occupying land through informal processes and building their homes through some kind of self-help construction. After all, this process has been described extensively in one Latin American city after another, and indeed, throughout most parts of the so-called Third World. In fact, the answer is by no means so straightforward, for many poor households rent or share accommodation. It is this issue that is at the heart of this book. We are principally concerned with establishing which kinds of family own their homes, and which rent or share them. What are the main factors influencing their tenure choice?

Traditionally, most urban Mexicans have lived in rental housing. Indeed, it is only since the Second World War that there has been a massive expansion of self-help housing. And, if it is true that the mass of the Mexican urban population now lives in self-help settlements, a substantial proportion of the occupants do not own their home. Depending on the city concerned, up to half the population live in homes belonging to other people. Many live in rental accommodation; others share homes with kin; some young adults continue to live with parents; some families look after homes for neighbours or friends.

If there was a pronounced shift towards owner-occupation after 1950, that tendency may now have been reversed. Indeed, there are good reasons for suspecting that the proportion of Mexican families occupying their own self-help home has declined during the 1980s. A combination of falling real incomes, rising costs of land and materials, and changing state policies may well have halted a seemingly endless tide.

In practice, we know too little about the residential preferences of the Mexican poor, and the economic constraints which face them, to know the answer. We do not know whether the majority aspire to the ownership of a self-help home. We are uncertain whether they yearn for such a home but simply lack the resources to realise such a dream. We do not know whether as cities grow in size some among the poor are forced to reject the option of peripheral ownership simply because of the difficulty of getting to work in the city centre. For this reason we are principally concerned with the factors underlying tenure choice.

## THE DEBATE ABOUT LAND

This book is intended to link the issue of renting to a number of basic issues raised in the theoretical debate about the role of self-help housing. One of the issues which has been extensively discussed in the literature relates to the cost of land in third world cities; in general, most commentators argue that the cost of land is rising, thereby slowing the growth of owner-occupation (Angel *et al.*, 1983; Doebele, 1987; van der Linden, 1987; Baross, 1987; Geisse and Sabatini, 1982; Durand-Lasserve, 1986). This argument differs radically from earlier thinking: for many years it was assumed that self-help housing flourished in most third world cities because land was easy to obtain. In Africa and parts of Asia, customary forms of landholding allowed newcomers to obtain free plots on the fringe of the city (O'Connor, 1983; Peil and Sada, 1984; Stren, 1982). In Latin America and the Middle East, land was often obtained through the medium of the land invasion, a process frequently encouraged by government and opposition parties alike (Drakakis-Smith, 1976; El Kadi, 1988; Danielson and Keles, 1984; Ward, 1982). Indeed, it was in a Latin American city dominated by land invasions that the doyen of the self-help housing literature formed most of his ideas. The very fact that John Turner worked in Lima, rather than some other city, was vital in convincing him that low-income households could successfully build and consolidate their own homes. A series of governments in that city had encouraged the invasion of public-owned land as a way of accommodating a burgeoning population and indeed garnering their political support (Collier, 1976). Had Turner worked elsewhere it would have been obvious that access to land was much more of a problem in most other large Latin American cities (Gilbert, 1981).

Today, many writers on third world housing contend that the period of free access to urban land is over. As Baross (1983: 205) argues:

> The majority of people who came to the large cities in developing countries in the last two or three decades found or developed housing in popular settlements. It was an historical epoch of non-commercialised or cheap commercial land supply. ... people did not have to pay or paid very little...this era in many developing countries is drawing to an end.

Similarly, UNCHS (1984: 25) have pointed out that

> non-commercialised processes for supplying land are disappearing in many countries... the land market is becoming increasingly commercialised. Land, particularly in the cities, is being quickly transformed from a resource with a use-value to a commodity with a market value.

Access to free land has become much rarer for a variety of reasons. First, the combined forces of demographic growth and suburbanisation have simply used up most of the accessible land. Rapid urban growth has exhausted areas of land which were previously available to low-income housing: 'pockets of unused or underused land ... have long since disappeared. Accessible public land has been inundated' (Doebele, 1987: 14).

Second, a fully commercialised land market has been established in most cities. Low-density suburban development has led to increasing numbers of middle-class owners occupying peripheral land. This has meant that most owners, including the public sector, have become well aware of the market value of their land. They have come to realise that the process of urban land conversion is a highly profitable business. It is not only the large landowner who is involved in this business: in the low-income settlements land invaders frequently sell spare land or subdivide their own plots (Gilbert and Healey, 1985). Growing commercialisation has become a dominant characteristic of the Latin American land market.

Third, commercialisation has encouraged the process of land speculation. Many owners of peripheral land have kept their property out of the market until the price of land has risen. Within the built-up area, many individuals hold plots for speculative reasons. The result is that large areas of land are excluded from the market, with obvious effects on the purchase price (Kowarick, 1988). As Trivelli (1986: 105) argues:

> The retention of urban land for speculative purposes, waiting for the general growth of urban areas to increase the value of land, is a common phenomenon in most cities. This speculative phenomenon has been driven to an extreme in the case of Brazil, where vacant lots represent one-third of the building space of Brazilian cities.

A similar process has also been documented in Santiago, Quito and Lima (ibid.).

Fourth, the search for profits through land speculation and real-estate development has led to much more sophisticated links developing between the real-estate, construction, and financial sectors. 'With economic development and city growth, finance and construction capital generally becomes progressively more important in the real-estate sector, and land revenues become more dependent on investments stemming from these sources' (Geisse and Sabatini, 1982: 162). Similarly, UNCHS (1984: 27) observe that

land and housing markets are becoming integrated and monopolised. There are indications that land development, housing development and housing finance are becoming increasingly integrated in large corporations with substantial land reserves, managerial capacities and access to short-term and long-term finance.

The operation of these large organisations makes it more difficult for small-scale operators to find land for low-income people. There is certainly evidence of this occurring in Bogotá where over the past fifteen years a handful of major companies has been developing middle-class estates on land that was once deemed fit only for low-income groups (Gilbert and Ward, 1985: 116).

Fifth, commercialisation has been further encouraged by the action of the state. In particular, governments have been much more prepared in recent years to upgrade irregular settlement and even to grant full legal title. As McAuslan (1985: 32) notes: 'Country after country has abandoned its neutral role in private market transactions and adopted regulations'. Clearly, this has brought benefits to many occupiers, particularly when it has given security of tenure to those living in fear of eviction. It has also led to major service and infrastructural improvements (Gilbert and van der Linden, 1987). But, the trend has also brought complications. 'Regularization causes an additional source of cost to illegal settlements. ...land is often bought or expropriated by the state and then sold to occupants' (Trivelli, 1986: 117). Similarly:

regularisation or legalisation can be a double-edged sword. For owners, it represents their formal incorporation into the official city, and the chance to realise what may be a dramatically increased asset. For tenants, or those unable to pay the additional taxes that usually follow, it may push them off the housing ladder altogether.

(Payne, 1989: 47)

Sixth, official intervention has sometimes brought additional problems for the poor. In particular, efforts to control urban sprawl have reduced the supply of land and particularly that available to low-income groups:

public authorities have taken strong measures to limit urban expansion into fertile agricultural areas, in efforts to preserve them for food production, for open space and for the control of pollution. Where such actions have been successful... [it has become] more difficult for low-income and disadvantaged groups to compete for sites ... sometimes pushing them to distant fringe villages.

(UNCHS, 1984: 25)

Finally, access to land has been hindered by the sheer physical growth of the city and has been worsened, in places, by the inadequate development of transport services (Kowarick, 1988). In many cities, peripheral land is now very far from the city centre. While jobs in factories or urban sub-centres may still be accessible, work in the central city is only available to those prepared to spend several hours commuting.

According to the literature, the result of these processes has been a rapid rise in land prices. According to UNCHS (1984: 26) 'land prices ... over long periods have tended to increase more rapidly than consumer price indices', and, according to Trivelli (1986: 103), 'land prices grow at a much faster rate than salaries'. The latter cites a series of sources which have documented sometimes spectacular rises in land prices. Even if few of these sources are methodologically very sound, there seems little doubt that prices have risen in real terms in most Latin American cities. Certainly, by far the most reliable survey of land values in any city, the World Bank study of Bogotá, shows that prices on the urban fringe are rising rapidly for all income groups, both in the formal and the informal market (Mohan and Villamizar, 1982; Carroll, 1980).

Despite this general consensus, a few words of warning are in order. First, it is by no means certain that land prices are bound to rise perpetually. After all, circumstances in Latin America have recently changed. While the 1960s and 1970s were a period of rapid demographic and economic growth, the 1980s have seen both a slowing in population expansion and a dramatic decline in economic prosperity. Certainly, real land and housing prices remained steady in Caracas between 1983 and 1986, and may well fall as a result of the acute crisis of 1989 (Gilbert, 1989a: 162). The economic recession is bound to slow increases in land values in most parts of Latin America, except where there is rapid inflation.

Second, if the price of peripheral land is rising rapidly as a result of the rural-urban conversion process, it is by no means certain that the cost of land within the already developed area will increase so dramatically. Certainly, land values in central Bogotá fell between 1955 and 1977 (Mohan and Villamizar, 1982) and evidence from low-income settlements in both Bogotá and Mexico City shows that real prices begin to fall after a couple of years (Gilbert and Ward, 1985: 113).

Third, generalising about land is fraught with difficulty because the situation in every city is different. Not only does the state of the economy differ but also the rate and form of urban growth. For example, some cities, such as Managua, Valencia and Monterrey, still have plenty of land, whereas the topography of others, such as Caracas or Rio de Janeiro, makes land very much scarcer. Size of city, the degree of land concentration, and most important of all, the attitude of the state varies considerably. As Trivelli (1986: 101) warns us: 'It is always risky to generalise about Latin America because situations vary considerably from one region to another and from one country to another. This is particularly the case for the operation of the land market'.

In practice, therefore, land market behaviour is likely to vary considerably between cities. And, while it is likely that real land prices will rise during periods of economic and demographic growth, the poor will not be excluded from access to peripheral land in every city. Indeed, that issue will not be determined wholly by economics; equally important is the attitude of the state to the poor. Latin American governments intervene heavily in the land market, with many states using land not only to reward construction and real-estate companies, but also as a means of seeking votes and political support. For this reason, governments as diverse as Odria's populist military regime in Lima and Venezuela's democratic administrations since 1958 have encouraged land invasions (Collier, 1976; Gilbert and Healey, 1985; Ray, 1969). Since governments change, there seems to be little consistency in the land allocation process. For example, in São Paulo, where land invasions were a rare occurrence before 1985, a change of national government in 1985 suddenly led to a major increase in this form of land alienation (Taschner, 1988). In Managua, the dreadful earthquake of 1972 cleared a large area of land in the centre of the city, which the Sandinista government later made available to low-income settlers. Elsewhere the flow of free land may suddenly dry up. In Santiago, Chile, the wave of invasions that characterised the period from 1969 to 1973, was instantly reversed by the Pinochet regime (Cleaves, 1974; Necochea, 1987). The result of this variation is that the precise nature of land allocation methods can only be determined at the local level.

The implication is that the cost of land access will vary considerably between cities (Gilbert and Ward, 1985: 110). Different methods of acquisition, forms of tenure, levels of income and expectation of servicing are all ingredients in determining the final price. As a result, it is unwise to make too general a statement about land price trends and the cost of access to land. Reality has a regular habit of confounding even the soundest forecasts.

However, in so far as there is evidence that the cost of land is rising in a number of Latin American cities, what effect will this have on tenure? The not unreasonable assumption in much of the third world housing literature is that it will lead inevitably to fewer home-owners and more tenants. Doebele (1987: 16) writes that

> the realities of the future will more probably be a housing market in which a much larger proportion of the poor dwell in rental units, and for whom the hope of ownership of land and house will become increasingly remote.

Van der Linden (1987: 7) speaks of the 'stagnating bridgeheaders' – those people who, ten or twenty years ago, would have become owners in an autonomous settlement, but can no longer afford to do so. Similarly, Amis (1982: 9) has observed that, in the case of Nairobi, 'today's squatter is tomorrow's tenant'.

In practice, the structure of tenure in any city does not depend entirely on the availability and cost of land. While land is vitally important, tenure is the outcome of a much more complex process. First, land prices are not the only ingredient determining the cost of self-help housing. The cost also depends on the price of building materials and services. In so far as evidence from many Latin American cities suggests that building costs have generally risen in real terms as a result of monopoly control over the supply of basic commodities such as bricks, glass and cement, this rise is likely to have slowed the growth of owner-occupation. On the other hand, this is not inevitably the result. After all, owner-occupation expanded rapidly during the 1960s and 1970s at a time when the prices of building materials were already rising rapidly. It is by no means certain, therefore, how rising costs of land and materials affect tenure choice among the poor. Rising land prices in irregular settlements, for example, may not deter home ownership so much as lower the quality of housing. Thus, faced by higher prices, settlers may simply buy smaller plots, a tendency clearly evident in Bogotá and Mexico City in the 1970s (Gilbert and Ward, 1985). Another likely response is for settlers to buy plots in less-accessible or in worse-serviced settlements. A further possibility is that they will pay more for the same plot and take longer to build and consolidate their home.

Second, the tenure balance depends not on the absolute cost of ownership but on the relative costs of renting versus ownership. If there is a plentiful supply of cheap rental units, many families may choose to remain as tenants rather than suffering the responsibilities and difficulties of self-help construction. Where land is very cheap relative to rents, owner-occupation may become a much more attractive proposition. An influential factor in this relative cost equation is state policy. Where governments have introduced rent controls, tenants with secure tenure and cheap rents may be reluctant to move. Where governments encourage their supporters to occupy public land, the cost of ownership relative to renting may fall dramatically. The relative cost of ownership versus renting varies dramatically from society to society as a result of very different kinds of state policy (Gilbert, 1989b). Different states have different attitudes towards home-ownership and dif- ferent approaches to increasing the supply of rental housing. Above all else, this variation in state policy explains the differences in tenure balance between countries and cities (Harloe, 1985). For this reason alone, there is no clear relationship between per capita income and housing tenure; as Table 1.1 shows, national affluence certainly does not lead inevitably to widespread home ownership.

Fourth, the cost and availability of transportation is also a critical element influencing tenure. Nineteenth century British cities lacked a self-help housing movement in part because no cheap, convenient transportation was available. As a result, workers were forced to live close to their jobs (Stedman-Jones, 1971). It was only when cheap workers' trains were introduced in London in the 1880s, that conditions in the desperately overcrowded rental accommodation began to improve. By the end of the century, Wohl (1971: 33) contends, 'cheap transport had completely changed the habits and mobility of the working classes'. Even in Latin American cities, self-help ownership did not become widespread until transport encouraged suburban development. Buenos Aires was one of the first cities to experience a process of self-help ownership simply because the spread of the electric tram network permitted a housing alternative to the central *conventillos* (Scobie, 1974). Elsewhere, self-help ownership did not become the norm until regular bus services allowed workers to move quickly and cheaply to work; most urban Latin Americans lived in centrally located rental accommodation well into the 1950s. Even today, unless good transportation is available, it is uncertain whether every family will choose owner-occupation even when they can afford it. Given increasing distances and the deteriorating transport situation in many cities, many may decide that the balance of advantage between renting a home near the city centre and ownership on the periphery favours the tenant option.

Finally, even if households are forced to forsake ownership in the immediate future, it is not certain that they will become tenants. There are alternative forms of tenure available. Some, for example, will share with kin, a phenomenon that has become increasingly apparent in Santiago in recent years (Necochea, 1987) and which has long been common in Mexico City (Sudra, 1976; Gilbert and Ward, 1985; Varley, 1985b). Rather than moving into rental accommodation or home-ownership, young adults will postpone their departure from the parental home. Newly-married couples will live with kin. The only obvious outcome of rising housing costs, therefore, is greater overcrowding and deteriorating conditions.

*Table 1.1* Owner occupation and per capita income, 1980

|  | *Year* | *% Owners* | *GNP per capita (1981 US $)* |
|---|---|---|---|
| Switzerland~ | 1980 | 30 | 17,430 |
| Sweden~ | 1981 | 57 | 14,870 |
| Norway~ | 1980 | 67 | 14,060 |
| West Germany~ | 1978 | 37 | 13,450 |
| Denmark~ | 1980 | 52 | 13,120 |
| USA~ | 1981 | 65 | 12,820 |
| France~ | 1978 | 47 | 12,190 |
| Belgium~ | 1981 | 61 | 11,920 |
| Netherlands~ | 1981 | 44 | 11,790 |
| Canada~ | 1978 | 62 | 11,400 |
| Australia~ | 1981 | 70 | 11,080 |
| Austria~ | 1981 | 50 | 10,210 |
| Japan* | 1980 | 60 | 10,080 |
| UK~ | 1981 | 59 | 9,110 |
| New Zealand~ | 1981 | 71 | 7,700 |
| Italy~ | 1981 | 59 | 6,960 |
| Eire~ | 1981 | 74 | 5,230 |
| Israel~ | 1978 | 71 | 5,160 |
| Venezuela[+] | 1981 | 75 | 4,220 |
| Argentina[+] | 1980 | 68 | 2,250 |
| Mexico* | 1980 | 67 | 2,250 |
| Brazil[+] | 1980 | 62 | 2,220 |
| Korea* | 1980 | 58 | 1,700 |
| Ecuador[+] | 1982 | 67 | 1,180 |
| El Salvador* | 1978 | 57 | 650 |
| Pakistan* | 1980 | 78 | 350 |
| Sri Lanka* | 1982 | 69 | 300 |
| Bangladesh~ | 1981 | 90 | 140 |

*Sources*:  Percentage owners:  ~ Boleat (1985 : 470)
                          * UN (1985)
                          + UNECLA (1986).
      GNP per capita:  World Bank (1982).

## THE NATURE OF THE STATE

We have already made several allusions to the role of the state. This is in line with recent work which has demonstrated how state policy is the critical element conditioning housing tenure (Daunton, 1987; Harloe, 1985; Kemp, 1987). Certainly, the housing issue can hardly be divorced from the wider political arena in Latin America where the literature has long since demonstrated how the state has manipulated low-income populations for political purposes. The encouragement of land invasions, the provision of infrastructure and services, and the use of community action programmes as a means of fostering political support, have all been used by governments in a variety of Latin American cities (Cleaves, 1974; Collier, 1976; Cornelius, 1975; Nelson, 1979; Portes, 1979; Ray, 1969).

The state has paid careful attention to housing because poor living conditions have long threatened to be a source of political protest in Latin America. Indeed, it was the fear that the hordes of migrants moving to the cities would one day develop into a revolutionary force that was a principal motivation behind the huge Alliance for Progress housing programmes established in most of the region's major cities during the 1960s. In practice, housing has not played as great a role in political conflict as some had feared and others hoped for. Certainly, the number of social movements arising directly from the crisis of collective consumption has been far fewer than Castells (1977) originally predicted. It was his basic contention that as urbanisation proceeded and urban life became more complicated, the state would be drawn inexorably into political conflict. As the needs and demands of the urban population increased, only the state would be able to provide the necessary services and infrastructure required. It would either have to provide the services itself or arbitrate between the suppliers and the users of the 'means of collective consumption'. But, as the state would be unable, given the contradictions of capitalist development, to satisfy conflicting demands, its involvement would politicise debate and lead to urban protests. When channelled by socially-aware groups, these protests might well develop into social movements aimed at the radical restructuring of society.

In fact, there have been relatively few major protests sparked directly by the housing issue in Latin America. Perhaps, only in Monterrey in the 1970s, Mexico City after the 1985 earthquake, and São Paulo in the 1980s, has the crisis of collective consumption really played a role in stimulating major protests (Bennett, 1989; Pozas Garza, 1989; Kowarick, 1988; Slater, 1987). Most protests have been brief and, more often than not, have usually been linked to transportation. If the crisis of collective consumption has produced a rather muted effect, it is partly because the state has managed to improve services in most Latin American cites, and partly because politicians of every persuasion have been very effective in discouraging protest, either through patronage or through repression (Gilbert, 1987). Castells has recognised this situation and has modified his position considerably in later writing (Castells, 1983).

The very nature of the self-help housing process has also contributed to political placidity. Widespread home ownership has individualised what might otherwise have constituted a more community-wide struggle. Not only has owner-occupation encouraged individual households to work hard to improve their own home, but it has kept them very busy in the process. It has also discouraged protest in case the state acts against their settlement and threatens their lack of legal status. And, while self-help communities have frequently pressured the authorities for land titles, water, electricity and other services, those demands have been relatively easily satisfied (Gilbert and Ward, 1985). In general, there can be little doubt that home ownership has helped stabilise Latin American cities. Indeed, within the Latin American political calculus, the ideology of home ownership has been very important. Few states have failed to recognise the advantages to social stability of increased access to home ownership. This realisation has been at the root, not only of middle-class housing subsidies, but also in the state permitting the proliferation of irregular settlement. As President Mariano Ospina of Colombia long ago recognised

> the day a citizen becomes the owner of a house, when he realises that the walls protect his wife and children and that those walls will protect them when he is dead, he is totally transformed and becomes at one with society.
>
> (cited in Laun, 1977: 311)

There is also some reason to believe that the division of low-income populations into owners and tenants has helped to divide those populations politically. Certainly, settlements of owners are more active than those with many tenants; the evidence shows that owners are much more active in petitioning for settlement improvements (Edwards, 1981; Gilbert and Ward, 1985; Nelson, 1979). Not only are tenants more recent arrivals, often coming to a settlement after the main battles have been fought, but they also have a much less obvious stake in securing improvements. They may even be involved in a conflict of interest in so far as settlement improvements may lead to substantial rent increases (Gilbert, 1989b).

What has also been surprising is that tenants have not been more involved in political struggle on behalf of their own housing 'class'. While there is some evidence of tenant struggle organised by the left in Bogotá, and considerable pressure applied on the state recently by the inner city tenant population of Mexico City, such activity has been remarkably uncommon in most Latin American cities (Massolo, 1986; Nelson, 1979). Possibly the difficulty of organising associations among highly transient tenant populations has been the main factor. Perhaps, too, the rhetoric and sometimes the presence of rent controls has helped reduce the level of protest. Or, perhaps, the aspiration of many tenants to become home owners in the not so distant future, has muted incipient protest.

Certainly, any kind of urban protest was made difficult during the 1970s by the rise of authoritarian regimes in most Latin American countries. Levels of repression rose under military regimes in Argentina, Brazil and Chile, and even in supposedly democratic countries such as Colombia and Mexico. So long as military leaders remained in power, open political protest was difficult. Gradually, however, as those authoritarian structures began to crumble, the openings for political action increased and the state began to re-establish more typical forms of political mediation. In general, this involved a delicate balance between stimulating continued capital accumulation and maintaining social stability. The state has been required to maintain a position of 'relative autonomy' from the dominant social classes, taking independent action when necessary in order to maintain social harmony. In order to perform this role, state policies have fluctuated from the extremely repressive to the highly populist. The poor have sometimes been rewarded with the offer of land or services; sometimes their street demonstrations or land invasions have been met by police brutality. Housing policy can only be understood in the context of this wider political process.

## THE ISSUE OF TENURE

Within these general debates the question of tenure has been rather neglected in Latin America, and indeed for most parts of the third world. Rather more has been written about nineteenth-century British cities than about renting in all of twentieth-century Latin America. Nor has the problem of accommodating large numbers of tenants figured in most national plans or in most housing policies. In Colombia, for example, while housing has often figured prominently in recent national plans, rental housing has never been discussed. In Mexico, national government agencies have invested heavily in housing programmes during the past couple of decades, but virtually all construction has been intended for ownership; rental housing has received little in the way of funding. In most parts of the third world, tenants and landlords have become 'invisible' (Grennel, 1972).

It is interesting to reflect on why this has been the case. First, self-help housing has come to dominate the political agenda, partly because the dramatic growth of this form of housing has clearly worried many politicians and administrators. The thought that hundreds of thousands of people were living in flimsy accommodation on the edge of their major cities generated fears about the threat to public order and health. But, more recently, self-help housing has also dominated the political agenda for the opposite reason. As a result of the arguments of writers such as Abrams (1964), Mangin (1967) and Turner (1967; 1968), self-help housing has been viewed as a possible panacea for the housing crisis. Self-help home ownership has been viewed as a means of pacifying the poor and of providing a cheap method of increasing the size and sometimes the quality of the housing stock. The 'slums of hope' were increasingly viewed as a potentially fruitful field for government involvement.

Second, it is clear that most governments were unenthusiastic about dealing with the problems of the central city 'slums of despair'. Such areas were often in an advanced state of physical decay and they were also thought to contain a high proportion of the families with severe economic problems. In addition, many of the occupants were newly-arrived migrants who many governments still hoped would return to their homes in the countryside. If this were not enough reason to neglect rental housing, undesirable political side-effects might ensue from any attempt to tackle the rental problem. Attempts to improve conditions in central areas might come into conflict with major real-estate interests; in places, such groups were simply too powerful to take on. Elsewhere, antipathy to the 'rentier' class discouraged governments from introducing programmes which might increase landlord profits rather than improve the position of the tenants. It is only very recently that there have been belated signs of a shift in attitude. International attention has once again begun to focus on the rental housing market. Within the World Bank, several officials have begun to recognise that rental housing programmes offer major opportunities for improving living standards (World Bank, 1980; Keare and Parris, 1982; Lemer, 1987; Mayo, 1985). The United Nations Centre for Human Settlements has been commissioning reviews of rental housing policy and individual governments have begun to introduce a rental component into their housing policies; national governments in Indonesia and Mexico have recently introduced rental housing programmes (Gilbert, 1989b; UNCHS/IHS, 1989; ISS, 1989). While this shift in approach will take a long time before it is translated into real action on the ground, the movement has undoubtedly begun.

# 2 Research strategy and a brief guide to Mexico

In the first section of this chapter, we summarise the main issues that we have attempted to tackle in the book. In the second section, we discuss the main elements of our research method. Finally, we include a brief résumé of Mexican economy, politics and society. The last section is justified in so far as many readers will be unfamiliar with Mexico and will otherwise be mystified by references made later in the book to institutions such as the PRI or the *ejido*. Needless to say, Mexicans and Mexicanists can skip this section and we request those who forego this invitation to be patient with our attempt to summarise the key features of Mexico in only a few pages.

## MAIN RESEARCH QUESTIONS

The basic aim of this book is to understand better the dynamics of the rental housing market. Here we are concerned with a whole series of broad questions which currently have few answers. On the basis of a wide literature review and on detailed data collected in two large Mexican cities, Guadalajara and Puebla, we ask: Who are the landlords and who are the tenants? What are the chances of most tenants and non-owners becoming owner-occupiers? What determines the proportion of a city's population that owns or rents accommodation? How is the rental housing sector organised? What is the role of public intervention in stimulating or depressing the rental housing sector? These are important questions at a time when the economic crisis, combined with a continuing tendency for land and house prices to rise, is generating increasing pressure on the low-income population in its search for housing. At a time when most governments now accept the inevitability of self-help housing solutions, there are questions about the continued viability of self-help in many Latin American cities. Deteriorating incomes among many of the middle classes are complicating the move to ownership throughout the region. As a result, several governments have begun to consider their policies towards rental housing.

After many years of neglect, rental housing is now reappearing on the political agenda.

Since we know so little we have attempted to tackle a wide range of issues. The major questions which we have sought to address are as follows:

## Residential tenure structure

It is clear that households do not choose accommodation in an unconstrained environment. Income levels relative to the price of housing determine the range of the options open to all but the richest families. But some high-income cities are dominated by owner-occupation whereas others have a predominance of rental accommodation. Why does the tenure structure of cities differ both across space and time? Is the structure dependent upon the interrelationship of demographic growth, housing supply and level of prosperity? Is the tenure structure a cultural feature with little basis in economic or demographic realities? In countries such as Mexico, what scope is given to self-help housing to increase levels of home-ownership? Can households in poor cities even afford to live in self-help homes? On the basis of both international comparison and Mexican data we consider different answers to these broad questions.

## The production of rental housing

In the past, the principal form of accommodation for the poor was rental housing located in the centres of the major cities. This was characteristic of most cities in the developed and less-developed world alike. Tenants could obtain accommodation because converted and purpose-built housing was available for rent. Until the early decades of the twentieth century, and in places later still, investment in rental property was a viable business. With the development of new channels for savings, with the introduction of government controls over rents and tenure, and with the emergence of suburban housing as a principal form of capital circulation, the profitability of rental housing has generally declined. In Latin America there has been a general decline in the proportions of urban homes for rent. None the less, in most cities the absolute numbers of rental housing units have increased, principally because self-help home-owners have expanded their property to accommodate other families. And yet we know very little about the dynamics of this process. We have little or no knowledge about the nature of these new landlords or the process of renting accommodation. To what extent is rental housing in Latin American cities a profitable activity? Under what circumstances should it be encouraged and what might governments do to stimulate more investment in rental housing?

## Residential trajectories and access to ownership

A principal hypothesis of this research is that non-owners are a residual population excluded from home-ownership by the expense of buying a house or even a plot of land. While some tenants undoubtedly choose to rent when they could afford to buy, the majority are earning too little to afford their own home or have accumulated too few savings. Partial evidence suggests that most families in Latin American cities aspire to home-ownership; certainly dominant ideology and economic realities underpin this aspiration. But, we have little firm evidence as to the strength with which poor people feel the need to own and the priorities that they place on home-ownership compared with alternative forms of expenditure or investment. We also know rather too little about the residential trajectories of different kinds of family. While we know that many families move from sharing with kin, to rental accommodation and then into home-ownership, we know rather little about the parameters of this trajectory. How has the current economic recession affected the eventual move into home-ownership? How have different kinds of government policy affected the viability of the sharer-renter-owner transition?

## Characteristics of landlords

Rental accommodation may be provided in large units by finance companies and affluent landlords or in many small units by owner-occupiers letting rooms in their own homes. There are clearly many intermediate levels of landlordism. In fact, we know very little about who lets accommodation, their aims, the profitability of their investment and their position in society. In most Latin American cities it appears that renting is now principally the preserve of small-scale owners. The current research examines the nature of landlords in Guadalajara and Puebla.

Not only do we know little about the economic and social characteristics of landlords, we are also remarkably ignorant of their political organisation and power. It is now widely recognised that it was the lack of political power that helped destroy the landlord class in Britain. How effective are landlords in Mexican cities in terms of influencing government policy?

## Relations between landlords and tenants

Do tenants have contracts? How long can they delay paying rent? What safeguards do they have against eviction? How do landlords choose between different kinds of tenant? How do they advertise vacancies, go about evicting bad tenants, and determine rent levels? To what extent does the state try to intervene in the landlord-tenant relationship and what are the effects of such intervention? None of these issues have been studied systematically in Latin American cities; the current work provides a first step towards better understanding of these questions.

## Government intervention

State policy is a critical ingredient in the changing housing situation in any city. Clearly, legislation on rent levels and mortgage relief, the nature of government building programmes, and the incentives available to private sector construction activity, are vital issues in determining housing conditions. But so too are other kinds of state policy. How does the state react to land invasions or other forms of illegality and irregularity in the land supply? What is the attitude to government subsidies for the poor? How do governments manage inflation and how do their policies impinge on the incomes of the poor? General economic and social policies are clearly highly influential in the housing equation. Given this multivariate equation, policies towards housing are less than easy to devise. A key issue posed in this book is how governments can increase investment going into rental housing.

## RESEARCH STRATEGY

In the preface we emphasised that this study forms part of a large endeavour to understand rental housing markets in Latin American cities. Within this context, Mexican experience is of obvious importance. Not only does the country contain some of the region's largest cities, but the proportion of tenants and other non-owners is relatively high. In this sense the choice of Mexico is unproblematic.

Our decision to work in two Mexican cities rather than one is justified by our belief in the value of a comparative approach. Past experience has taught us that comparative work complicates the task of research but has the advantage of discouraging overgeneralisation. All too often, academics are wont to draw conclusions on the basis of experience in a single city, making generalisations about the whole of Latin America or sometimes, even, about all 'third world' cities. The particular advantage, and difficulty, of comparative research is that it constantly reminds the researcher that every city is different. The regular observation that what happens in one city does not automatically happen in another forces the researcher to think harder about underlying processes. Certainly, a comparative approach is not without its own kinds of problems, but in this respect it does impose its own intellectual rigour.

The choice of where in Mexico to work was determined on the basis of several decisions. First, we decided not to work in Mexico City, partly because of its sheer size and complexity, and partly because of the volume of work that had already been carried out on housing issues in that city. Second, we wished to work in large cities rather than small ones so that we could attempt to relate our findings to conditions in other Latin American metropolises. Third, we wanted to choose cities with large numbers of tenants rather than those with a large majority of home owners. While a case could have been made for comparing one city with many tenants with another with few, our lack of information about the nature of the rental housing market persuaded us to work in two cities with many tenants. Since Guadalajara and Puebla were not only the country's second and fourth largest cities, but also contained, in 1970, the highest levels of non-ownership among cities with more than 200,000 inhabitants, these were the decisive factors in their selection (see Table 3.1). Finally, Guadalajara and Puebla were interesting in so far as they constituted very different kinds of cities. Guadalajara has a dynamic economy which dominates a vast region of its own, whereas Puebla, although growing equally rapidly, lies in the heavy shadow of Mexico City (see Figure 2.1). Demographically, Guadalajara's population has little in the way of ethnic minorities, whereas Puebla has many inhabitants who were born in Indian areas. Politically, both are conservative cities with a strong religious tradition, but Puebla continues to support the dominant political party, whilst Guadalajara is much more independent.

The bulk of the primary research material was collected between August 1985 and September 1986 with short follow-up visits being made in 1987 and 1989. The information in this book comes from a range of sources. First, we reviewed both the Mexican and the general literature on rental housing, making a deliberate attempt to try to integrate some of the findings of first and third world research. Second, we read the main newspapers in Guadalajara and Puebla and the principal weekly and monthly magazines in Mexico City for the period since 1975. Third, we interviewed government officials, property managers, leaders of tenant associations, academics and other people familiar with the housing scene in the two cities. We also talked to the major housing and planning institutions in Mexico City and to numerous academics in the Autonomous Metropolitan University (UAM), the National Autonomous University (UNAM) and the Colegio de México. Finally, we carried out a major survey of owners and tenants in Guadalajara and Puebla as well as a small survey of landlords.

The 753 interviews with owners and tenants constitute our main source of original data. The survey was not intended to be representative of the whole population of each city, nor of the urban poor. We chose instead to

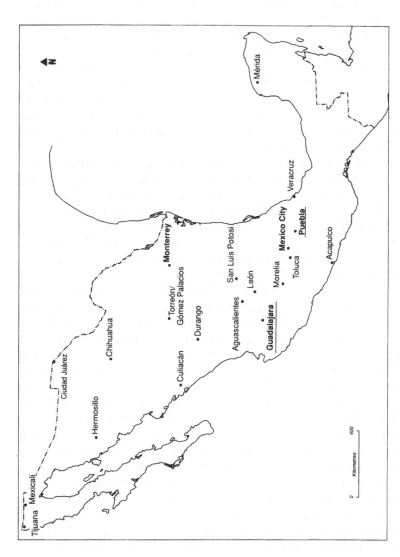

*Figure 2.1* **Mexico: Location of major cities**

work in three different kinds of settlement in each city. These kinds of settlement are found throughout urban Mexico and in practically every Latin American city as well. Crudely, they represent central areas with large rental housing units, well-consolidated self-help neighbourhoods, and newly-established peripheral settlements. By choosing a range of settlements we would find households at different stages of their residential history, thereby examining the validity of some of the principal theories about urban residential mobility. We would find newly-established tenant households as well as those that had been renting accommodation for many years. Similarly, we would find both long-established and newly-formed owner-occupier families. Different kinds of renting and ownership would also fall within our purview.

The first kind of settlement was intended to represent the central rental areas to be found in most large Latin American cities. Many tenants would live in single rooms, sharing communal services, in large, deteriorated *vecindades*.[1] These tenants, as well as those renting whole apartments or houses, would be drawn from the older inhabitants of the city, and many of them would work near by. The two settlements, Central Camionera in Guadalajara and Analco in Puebla, were both located within 2 kilometres of the central cathedral (Figures 2.2 and 2.3).

The second kind of neighbourhood would be a well-consolidated self-help settlement and most houses would have access to essential services such as water and electricity. Renting would be well established within the settlement and there would be approximately equal numbers of owners and tenants. In Guadalajara, Agustín Yañez was selected in the east of the city, 7 kilometres from the cathedral (Figure 2.2). It had been founded by illegal subdivision in the early 1950s. In Puebla, we chose Veinte de Noviembre, a settlement located about 4 kilometres to the north-west of the cathedral and founded by illegal subdivision in the late 1940s (Figure 2.3).

The third kind of settlement was to represent a new, self-help community located on the fringe of the city. The neighbourhood would still have empty plots, unconsolidated houses, and little in the way of infrastructure and services; it would contain few tenant households. Interviews in this neighbourhood would garner evidence of how households in self-help communities had come to be owners. Since the communities would have been founded recently, they would provide up-to-date information about the difficulty of acquiring land in each city. In Guadalajara, this kind of community was represented by Buenos Aires, a settlement founded on *ejido* land[2] some 9 kilometres south of the cathedral (Figure 2.2). The settlement had been founded in 1979 and most occupants had arrived between 1981 and 1983; new settlers were still arriving. The settlement lacked piped water and sewerage and some homes lacked electricity, which

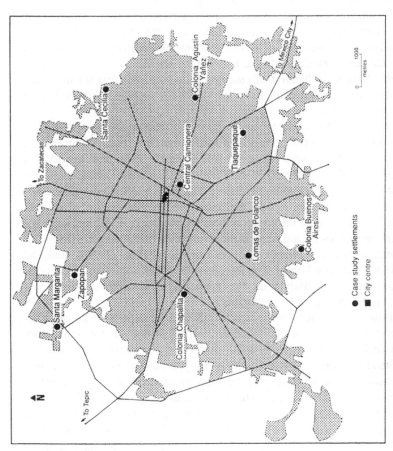

*Figure 2.2* **Guadalajara: Location of case-study settlements**

*Source*: Figure 5.3

was being supplied illegally during our research. In Puebla, the self-help settlement chosen was El Salvador (Figure 2.3). Founded in 1981 on *ejido* lands, the settlement was located 6 kilometres east of the cathedral. Many of its plots were still uninhabited and the majority of its inhabitants had arrived between 1982 and 1985. Water was obtained through the perforation of wells and there was no drainage system.

Only tenants were interviewed in the central neighbourhoods and only owner-occupiers in the peripheral settlement. In the consolidated self-help settlement similar numbers of tenants and owners were visited. Households were selected at random from a list of families identified in a previous household count. A handful of students from local universities were employed to assist in conducting interviews, with the average interview lasting some twenty minutes.

Interviews were conducted with either the man or the woman householder and data were collected on the whole household since its formation. Information was obtained about the place of birth, migration history and employment of both the male and female householder, and about the household size and structure, income, residential history, quality of accommodation, and attitudes to past and future residential choice.[3] Different kinds of supplementary information were sought from owners and tenants.[4] Owner-occupiers were questioned about how they had managed to mobilise the funds to become owners and about the acquisition of their land and construction of their homes. The tenants were asked about their landlords, previous rental accommodation, the conditions of their tenancy, and their interest in becoming owner-occupiers.

Follow-up interviews were conducted with a number of households. Sixteen tenants in a *vecindad* in Analco were visited and a detailed history of the *vecindad* was made. In-depth interviews were also made with the inhabitants of another young settlement on *ejido* lands in the east of Puebla.

In addition, a separate survey was conducted among landlords. Unfortunately, this survey could not be based on a rigorous sample because it proved impossible to identify every landlord. Resident landlords were no problem but non-resident owners were often elusive. Tenants were the main source of information about the landlords but in Analco and particularly in Central Camionera many did not know the address, or, sometimes, even the name of their landlord. Landlords were not interviewed unless they were known to have more than one tenant, because we did not wish to cause any difficulties for tenants. This procedure limited our choice of landlords, particularly in Agustín Yañez and Central Camionera. Eventually, ten landlords were interviewed in Guadalajara and thirty-seven in Puebla. Qualitative rather than quantitative information was sought in the interviews with the landlords, some of whom talked with us for an hour or more.

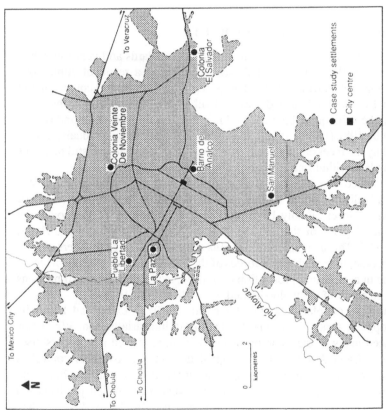

*Figure 2.3* **Puebla: Location of case-study settlements**

*Source:* **Figure 5.4**

They were asked about how they became landlords, about the nature of their property, how they determined appropriate rent levels, how they chose tenants, how and why they went about eviction, how they managed their properties, and about the general economic, political and legal environment which faced them. Interviews were also conducted with a dozen representatives of firms managing rental property. Most of these were located in Puebla where a higher proportion of landlords appeared to use such firms.

## MEXICAN ECONOMY AND SOCIETY

Mexico is relatively rich by Latin American standards and very rich when compared with most parts of Africa or Asia. As a result of continuous economic growth from 1940, it had become the fifth richest country in the region by 1960. When it discovered oil in large quantities in the early 1970s, the subsequent boom turned it into Latin America's second richest nation by 1982 (IADB, 1988). This elevated position was gradually undermined by the recession that began that year, although Mexico still manages fourth place within Latin America (World Bank, 1989).

If Mexico is still a relatively affluent country, that wealth is very unequally distributed. In 1985, it is estimated that the poorest 40 per cent of households received only 13 per cent of total household income compared with the 51 per cent share of the richest one-fifth (Hernández and Parás, 1988: 971). This unequal pattern means that there are large numbers of very poor people; indeed, the joint forces of the economic recession and a recent decline in agricultural production currently place an estimated 40 millions in the category of the malnourished (Cornelius *et al.*, 1989: 5).

Until the petroleum boom of the mid-1970s, Mexico had based its development strategy on import-substituting industrialisation. While this produced many of the same problems that characterised the process elsewhere in Latin America: technological dependence, overseas control, a balance of payments deficit, capital intensity, and urban bias, the experience was not unsuccessful. At the very least, it had produced an average growth rate of over 6 per cent between 1940 and 1960 and had created a major manufacturing sector. Living standards had risen generally and a large middle-class had developed in the cities. The poor did not share equitably in economic growth but after a period of decline during the 1940s, real manufacturing wages increased by two and a half times between 1950 and 1975 (Bortz, 1984: 351–2).

The cities gained most of the benefits to be derived from the import-substitution process. As a result, huge numbers of migrants moved into the cities, gradually turning a rural society into an urban nation. Between 1950 and 1980, the number of people living in the largest twenty-five cities increased from 5.7 to 26.1 million, the proportion of Mexicans living in the countryside falling from 57 per cent to 34 per cent in those thirty years. Despite the pace of urban growth, the authorities managed to cope with that expansion. They provided most of the urban population with electricity and schools, and a substantial proportion with health care, water and sanitation (IADB, 1988; Ward, 1986: 88).

Admittedly the burgeoning urban population was accommodated largely by their own efforts. While the middle class increasingly bought houses subsidised by the public sector, the majority of poor families lived in self-help settlements. In this respect, Mexico was an archetypal Latin American society, its cities carefully divided into discrete social areas, affluent and attractive suburbs coexisting, albeit at a distance, with vast areas of poorly serviced and unattractive low-income settlement. The state helped to segregate rich from poor, implementing urban plans in the more affluent areas and generally turning a blind eye to the irregularity of most self-help communities (Garza and Schteingart, 1978; Gilbert and Ward, 1985; Varley, 1985b).

Of course, like other countries in Latin America, Mexico has long faced a terrible housing problem. Even if conditions are vastly superior to those found in most African or Asian countries (O'Connor, 1983; Peil and Sada, 1984; Gilbert and Gugler, 1982), overcrowding is widespread. In 1980, the average number of persons per room in Mexico as a whole was 2.3 and more than one-third of urban households lived at densities of more than two persons per room (IADB, 1988: 59; *Censo General de Población*, 1980). And, while service provision has tended to improve, it is still far from good. Although electricity provision is virtually complete in the urban areas, the homes of almost one-quarter of urban households lack access to the main sewerage system (IADB, 1988: 27, 60–3).

The Mexican state has certainly made an attempt to improve living conditions in the cities. At the same time, its efforts have reflected its dominant ideology and political approach. The dominant goal has been to maintain political stability and to sustain the process of economic development. The political system has been dedicated to those goals, if necessary at the cost of other subsidiary values such as human rights, democracy and equality. Until the early 1980s, it achieved its goal with some success. With the exception of the 1968 massacre, the monolithic Institutional Revolutionary Party, the PRI, had managed to run the country without recourse to excessive violence. It had provided every president since 1929 and only in 1988 did its candidate come under any kind of serious electoral threat. While it ran the country in an authoritarian manner, it avoided the extremes of violence and repression that were common in the southern cone countries. While it regularly suppressed outright opposition and rigged elections, it allowed something in the way of consultation and involvement. Indeed, its whole *modus operandi* was geared to corporatist politics; the three main organisations which make up the PRI represent the peasants, the workers, and the so-called popular sector. The PRI maintained some measure of harmony by recruiting people from every rank of society into the government and party bureaucracies. Indeed, its ability to reward its supporters and to coopt or buy off opposition figures where necessary was the key to continued political stability. If Mexicans played the party game, they would be rewarded. Seemingly the only crime within the PRI was disloyalty; corruption was permitted but not criticism.

As a result, the Mexican people have long viewed the state as being 'distant, elitist and self-serving' (Cornelius *et al.*, 1989). Since stability has been more important than honesty or fairness, much has been done in the name of the PRI that did not improve the state of Mexican society. The unions and the police were known to be corrupt but so long as they remained loyal, nothing much was done to control their excesses. And yet, Mexico's government cannot be accused of incompetence. It has increasingly recruited personnel with high levels of technical competence; indeed there has been increasing unease among traditional politicians that they were being excluded from key areas of decision-making. Until the oil boom, it managed economic growth efficiently and its recent performance in controlling inflation and maintaining social harmony is a masterful display of high-wire politics. It has rarely resorted to outright bloodshed, the student massacre in 1968 being the most notorious exception, and when emergencies have threatened it has often performed with surprising alacrity. Its response to the effects of the 1985 earthquake, for example, has been widely praised. It is this general approach of cautious flexibility in search of stability that has arguably served Mexico reasonably well in the past. It is only in the 1980s that the system seems to have shown severe signs of strain.

Ironically, the principal catalyst of Mexico's current problems was the discovery of major new petroleum reserves in the 1970s. Given the high level of oil prices, increased production brought a vast influx of foreign earnings and greatly boosted the federal government's budget. Unfortunately, this encouraged a process of what in hindsight was an excessively rapid rate of economic expansion. Economic investment was financed through foreign loans, not an unwise policy at a time when oil prices were high and rates of world inflation were substantially reducing the real cost of borrowing. However, when interest rates rose in the United States and Europe in the early 1980s and the price of Mexico's petrol exports declined, the scene was set for the current crisis. By 1987 its foreign debt was US $1261 per capita, a little more than half its per capita income. The total debt of US $ 102 billions in 1987 was the equivalent to almost four times the annual income from foreign trade.

*Table 2.1* Mexico: principal economic and social indicators

| | | | |
|---|---|---|---|
| Population (1988) (millions) | | 83.0 | (a) |
| Population growth rate | (1970–75) | 3.2 | (a) |
| | (1980–95) | 2.4 | (a) |
| *Main economic indicators* | | | |
| Gross national product per capita (1987) (US $) | | 1,830.0 | (d) |
| Annual growth rate | (1970–80) | 3.5 | (a) |
| | (1981–88) | -1.1 | (a) |
| Total disbursed external debt (1987) (US $ billions) | | 102.4 | (a) |
| Debt/exports of goods and services (1987) | | 372.0 | (a) |
| Annual rise in consumer prices (1982–88) (per cent) | | 91.1 | (a) |
| Annual change in real urban minimum wage (1980–88) | | -6.7 | (a) |
| *Structure of production and employment* | | | |
| Agriculture in GDP (1985) (per cent) | | 8.6 | (b) |
| Manufacturing (per cent) | | 21.8 | (b) |
| Other industry and mining (per cent) | | 10.6 | (b) |
| Commerce and services (per cent) | | 59.0 | (b) |
| Labour force in agriculture (per cent) | | 25.8 | (c) |
| *Social indicators* | | | |
| Adult literacy (1985) (per cent) | | 90.3 | (a) |
| Life expectancy at birth (1980–85) (years) | | 67.4 | (a) |
| Infant mortality (1985) (per thousand) | | 49.9 | (a) |
| Population per doctor | | 1,037.0 | (a) |
| *Urban population* | | | |
| Proportion of total population (1980) | | 66.3 | (c) |
| Proportion living in cities with more than 100,000 inhabitants | (1950) | 28.2 | (a) |
| | (1980) | 51.4 | (a) |
| Average annual growth rate | (1970–80) | | |
| | | 4.5 | (c) |

*Sources:*

(a) UNECLA (1989).
(b) Hernández and Parás (1988: 969).
(c) Mexico, INEGI (1989).
(d) World Bank (1989: 164).

The effects of the debt crisis and consequent government policies have been dramatic. Between 1982 and 1988, the gross domestic product per capita declined by 11 per cent (UNECLA, 1989). Inflation rose to unprecedented levels, rising by 159 per cent during 1987 and averaging 91 per cent between 1982 and 1988 (Banco de México, *Indice de Precios*). In an effort to pay interest on the debt and to reactivate and transform the economy, the government has adopted a dramatic combination of policies including liberalising imports, privatising large numbers of parastatal organisations, savagely cutting government expenditure and subsidies, and holding down the real minimum wage. The combination of policies has produced a major fall in real incomes, and despite the official figures, rising levels of unemployment. It is estimated that the real value of most Mexicans' wages has been roughly halved since 1982. If a recent survey in Mexico City suggests that household incomes have fallen more among the middle class than among the poor, it is only because employment participation rates rose among the very poor from 1.3 persons per household in June 1985 to 1.9 in February 1988 (Mexico, INC, 1989: 53).

Table 2.1 summarises Mexico's principal economic and social indicators.

# 3 Residential tenure in urban Mexico since 1940

The Mexican housing system was transformed after 1940 as a predominantly rural country became a nation of city dwellers and a major shift occurred in the pattern of urban residential tenure. A situation where the urban population was housed as tenants changed to one where there is a predominance of owners. In 1950, only three Mexican cities with more than 100,000 inhabitants (Culiacán, Matamoros and Mérida) had a majority of owner-occupied homes. By 1980 the transformation was almost complete. Table 3.1 shows that practically every large Mexican city had a predominance of owners, many cities with two-thirds of their households in that category.

This tenure shift accompanied the demographic and physical transformation of Mexican cities. For a start the urban population exploded both absolutely and relatively. During most of the century Mexico had remained a predominantly rural society. By 1960, however, a bare majority of the Mexican population was living in settlements with more than 2,500 inhabitants.[1] As a result of rapid falls in the death rate, most cities were growing rapidly: many at around 5 per cent per annum, their populations doubling every fourteen years. The physical area of these cities obviously increased, not only because of the exploding numbers of people, but also because of the technological and economic processes underlying that expansion. Small, high-density cities were turned into sprawling urban areas. The ostensible causes of this pattern of growth are the same as those which have been governing urban growth throughout the capitalist world. Demographic growth combined with transport and service improvements, allied with the development of suburbs and, in Mexico and other Latin American countries, the proliferation of self-help housing, have wrought this change.

*Table 3.1* Proportion of homes occupied by owners by city, 1950–80

|  | *1950* | *1960* | *1970* | *1980* |
|---|---|---|---|---|
| Mexico City* | 25.0 | 21.8 | 43.3 | 53.5 |
| Guadalajara** | 29.1 | 28.3 | 43.1 | 52.1 |
| Monterrey*** | 36.2 | 31.2 | 45.8 | 68.0 |
| Puebla | 20.8 | 16.1 | 38.7 | 47.5 |
| Ciudad Juárez | 32.7 | 31.8 | 50.0 | 57.6 |
| León | 46.0 | 40.1 | 58.2 | 62.7 |
| Mexicali | 54.1 | 46.5 | 62.0 | 64.3 |
| Torreón/Gómez Palacio**** | 44.6 | 40.4 | 50.2 | 66.5 |
| Tijuana | 49.1 | 40.1 | 52.4 | 51.7 |
| Culiacán | 58.3 | 53.8 | 71.1 | 79.1 |
| Chihuahua | 39.9 | 38.8 | 52.5 | 71.0 |
| San Luis Potosí | 44.7 | 36.6 | 56.7 | 62.0 |
| Mérida | 56.8 | 51.1 | 71.0 | 77.2 |
| Toluca | 61.2 | 44.2 | 64.4 | 62.6 |
| Acapulco | 61.9 | 48.3 | 65.1 | 69.7 |
| Veracruz | 39.1 | 28.5 | 44.3 | 48.3 |
| Aguascalientes | 44.0 | 35.6 | 50.8 | 55.8 |
| Morelia | 57.0 | 43.2 | 60.2 | 64.1 |
| Hermosillo | 51.5 | 46.8 | 64.4 | 71.3 |
| Durango | 49.3 | 49.0 | 60.1 | 67.9 |

*Sources:*  1950–70  Coulomb (1981: Table 1.20) elaborated from *Censo General de Población.*
1980  *Censo General de Población y Vivienda.*

*Notes:*  Table includes all cities with more than 200,000 inhabitants in 1970. Cities are listed in order of their population in that year (based on data for the *municipio*, even if the city only occupies a part of it, because in some years tenure data are not available by enumeration district).

*Figures for the main municipio only except for the following:*
* 1950 Federal District only.
  1960 includes Federal District, Naucalpan, Tlalnepantla and Ecatepec.
  1970 includes the above plus Atizapán de Zaragoza, Chimalhuacán, Coacalco, Cuautitlán, Huixquilucan, Netzahualcóytol, Los Reyes-La Paz, and Tultitlán.
  1980 includes the above, plus Cuautitlán-Izcalli.
** 1980 includes Tlaquepaque and Zapopan.
*** 1980 includes San Nicolás, Guadalupe, Garza García and Santa Catarina.
****1980 includes Ciudad Lerdo.

## TENURE PATTERNS

Most rural people in Mexico have always housed themselves through self-help construction. In contrast, throughout the nineteenth century and up to the 1940s, Mexico's cities were dominated by tenants. Rental housing was not only the typical form of working-class accommodation but was also home for most of the middle class. Only the elite owned their homes, usually in blocks close to the town's main square.

There seems to have been little real variation between cities. Reports of conditions in most Mexican cities are agreed that renting was dominant until the middle of this century. In Mexico City, Connolly (1982: 148) states that 'rented housing in *vecindades* was practically the only alternative available to the working class population'. More detail is provided by Perló (1979: 789–90) who says that

> until the beginnings of the thirties, the most important housing niche of Mexico City's working classes continued to be the rented housing of the central area. The state housing system was insignificant and the appearance of shacks ... had only begun to take place, in the interstices of the old housing system.

In other cities too, rental housing was the dominant pattern, even as late as 1960 (Table 3.1).

## WHAT CHANGED THE URBAN TENURE SITUATION?

A major factor leading to change was the population explosion in urban Mexico after 1940. Table 3.2 shows how steady rates of growth in the largest cities during the 1920s and 1930s more than doubled during the 1940s and continued at still higher levels during the 1960s and 1970s. The combined population of the twenty-five largest cities in the country grew by 2.4 million people during the 1940s, 4.8 millions during the 1950s, 6.4 millions during the 1960s, and 9.1 millions during the 1970s. The combined population of these cities increased from 17 per cent of Mexico's total population in 1940 to 39 per cent in 1980. The scale of this growth was such that some change in the housing situation was virtually inevitable.

During the 1920s and 1930s, most urban population growth was absorbed in the traditional way: more *vecindades* were created and where such investment was insufficient overcrowding increased. From 1940, however, the shift to self-help housing began to complement this process. The change was due to a number of factors. First, it was clear that the existing housing system was suffering major problems even before the huge increase in the numbers of people requiring accommodation. In some respects, rental housing was an inadequate housing solution as early as the first decades of the century. Certainly, there is plentiful evidence of tenant protest about housing conditions. Tenant complaints, and sometimes strikes, frequently forced the state to intervene (see Chapter Four). Poor sanitary conditions, high rents during periods of recession, overcrowding and unfair treatment by landlords were recurrent issues in the urban politics of the twentieth century. González Navarro (1974: 174-95) demonstrates that although rent strikes were relatively uncommon, occurring in Yucatán (in 1915), Veracruz (1922), Mazatlán (1935), and Tepic (1938), local authorities were often obliged to ameliorate rental housing conditions. They attempted to dissolve incipient protest through a combination of rent controls, sanitary improvements, and modifications to local housing regulations.

*Table 3.2* Urban growth in Mexico, 1900–80

|  | Population of 25 largest cities (thousands) | % of national population | % increase over previous Census |
|---|---|---|---|
| 1900 | 1,260 | 9.2 |  |
| 1910 | 1,561 | 10.3 | 23.9 |
| 1920 | 1,858 | 13.0 | 19.0 |
| 1930 | 2,529 | 15.3 | 36.1 |
| 1940 | 3,345 | 17.0 | 32.3 |
| 1950 | 5,706 | 22.2 | 70.6 |
| 1960 | 10,526 | 30.2 | 84.5 |
| 1970 | 16,919 | 34.5 | 60.7 |
| 1980 | 26,054 | 38.7 | 54.0 |

*Sources:*  1900–50  Unikel *et al.* (1976: I-A1) and own calculations.
1960–80  Negrete and Salazar (1986: 113–14) and own calculations.

Second, it is clear that the profitability of rental housing was already declining after 1940. There were numerous causes of this decline. One was the general poverty of most urban dwellers, which meant that new investment was only profitable when allied with overcrowding and poor housing conditions. However, this option was complicated, as far as landlords were concerned, by increasing intervention from the state. Different efforts to control rents and improve sanitary conditions were tried in different cities, but in general they tended to make the traditional pattern of investment in *vecindades* more difficult. Some states increased their demands on landlords to improve sanitary conditions and to increase the services available in the *vecindades*. Stricter building regulations were making it more difficult to build substandard housing on a legal basis. In Mexico City: 'Given that the construction of low-rent housing in accordance with the building regulations was clearly not a viable concern, the production of new *vecindades* in the central and more controlled areas of the city was decisively hindered' (Connolly, 1982: 149).

Perhaps the rudest shock came, however, when Mexico was embroiled in the Second World War (Gertz Manero, 1964: 39). As in other Latin American countries, this was the moment when rent controls came to be firmly established. While the legislation was taken up with different levels of enthusiasm in different parts of the Republic, there was little doubt from this time on that state policy towards rental housing would constitute a problem for most owners (see Chapter Four). This was certainly the pattern in the Federal District where the rent freeze continued until 1948, and indefinitely for existing contracts (Portillo, 1984: 38). If rent controls alone cannot explain the decline of rental housing, along with other forms of legislation, they were a significant factor in reducing the profitability of this kind of investment (Coulomb, 1981: 161; Coulomb, 1985b).

Third, there were other opportunities for investors opening up in Mexico. Rapid commercial and business expansion in the cities was leading to large rises in the value of centrally located land, making alternative uses of that land more attractive. In addition, a whole series of new investment opportunities were appearing in Mexico after 1940. Political stability allied to industrialisation meant that the economy was beginning to grow quickly. As capital and financial markets became more sophisticated, share-ownership became more common and the development of banks made indirect investment far easier. Home-ownership also began to make economic sense for those with sufficient savings. If ownership of *vecindad* housing had traditionally been the safe way to invest accumulated or inherited savings, new and more remunerative opportunities were now appearing.

Fourth, other policies of the state were influential in modifying the pattern of housing. Of critical importance was the attitude of the state to the occupation of land by the urban poor. Whether land is easily available to low-income groups on the periphery of the city is critical to the process of self-help housing construction. It is arguable that the high level of renting in the cities of Mexico was a consequence of the control exercised over peripheral land by a reduced number of *latifundistas*. Monopoly over land precluded self-help housing of the type practised in rural areas. Perló (1981), in fact, argues that it was the undermining of these monopolies that was the most significant change in the pattern of urban development in Mexico City. It was the agrarian reforms of Lázaro Cárdenas and the creation of *ejido* communities out of *haciendas* that was critical. 'From the

Cárdenas era (1934–40) on ... there were clear signs of a new housing system: the low-income subdivisions' (Perló, 1979: 790). The combination of new *ejidos*, the sale of peripheral land by owners scared by the prospect of agrarian reform, and the general tolerance of Cárdenas to land invasions were all influential in stimulating the process of self-help housing in Mexico City (Perló, 1981: 74). This thesis, however, has been criticised on various grounds: a large part of the *ejido* land around Mexico City had been created much earlier, in the 1920s; very little *ejido* land was actually urbanised during Cárdenas' presidency; and it is difficult to see Cárdenas' agrarian reform as a deliberate policy for making land available for housing uses (Varley, 1989a: 131–5; Cruz Rodríguez, 1982: 34). The creation of *ejidos* certainly encouraged owners to urbanise their lands in an attempt to avoid the reform, but this process was already taking place in the latter years of the Revolution. What can be said about the agrarian reform and housing developments in the 1940s is that the vast majority of the *ejido* land around Mexico City had already been created by that time (Varley, 1989a: 134). As the demand for housing grew, this land came to constitute a major opportunity for low-income housing development. Post-Cárdenas presidents, unsympathetic to the agrarian reform, would do little to impede the conversion of *ejido* land into self-help housing areas.

Fifth, shifts in the housing pattern of Mexican cities were only conceivable in the light of changing patterns of technology. From the 1930s on, the motor vehicle was entering Mexico in increasing numbers. In 1924 there were only 32,531 cars registered in the entire country; by 1930 there were almost twice that number, and by 1940 nearly three times as many (Mexico, INEGI, 1985). After 1950 the rate of increase quickened, the numbers rising from 173,000 in 1950, to 483,000 in 1960, 1.23 millions in 1970 and 4.26 millions in 1980 (ibid.). The expansion of car ownership, and the public bus system, together with the changes in technology that allowed electricity, water and sewerage to be provided over increasing areas, opened up the possibility of suburban development for every class. The upper and middle classes were rapidly shifting to a suburban life-style. Changing house styles and a stronger demand for services were increasing the desirability of suburban living. As a result, the elite moved increasingly from central locations to the more attractive areas of the periphery. The poor followed, but towards less desirable parts of the city and using a decidedly different building style.

In sum, growing numbers of people, a changing investment climate, a modified response by the state to housing and land issues, changing technological restrictions on urban growth, and changing tastes in accommodation and urban living were all producing a situation which encouraged the widespread shift from renting to legal or *de facto* home-ownership.

## DIFFERENCES BETWEEN CITIES

The potential for a shift from renting to ownership was apparent in Mexico during the late 1930s. Eventually, every city in the country would experience a substantial change in its tenure structure. However, this shift did not occur at the same pace or begin at the same time in the different cities of the Republic. Table 3.1 shows that by 1960 owners still constituted less than one-third of the households in three of the country's four largest cities.[2] Non-ownership was still increasing in many cities in the 1950s and it was only in Mérida and Culiacán that more than half of the households were owner-occupiers. Even by 1970 there were still several predominantly rental cities.[3]

Why did such marked differences exist between the cities? One possible explanation lies in the timing of cityward migration and consequent urban growth. Although every city would eventually resort to self-help processes, because it was the only way to accommodate the growing population, the shift from rental housing to self-help ownership might occur at different times. Prima facie, it seems reasonable to explain the differences in tenure shift in terms of the timing and pace of population growth in different cities. For while heavy in-migration was to affect most cities of the Republic, some experienced substantial growth well before others.

Between 1910 and 1940, most of the larger cities grew considerably, but few grew so rapidly that they could not accommodate the increasing population through the traditional pattern of rental housing. It was only in Mexico City, Torreón, Monterrey and, most dramatically, Ciudad Juárez, that the pace of expansion made this solution problematic. The population of Mexico City grew from 630,000 to 1.8 millions between 1910 and 1940; that of Torreón nearly quadrupled; Monterrey added more than 100,000 inhabitants. Some of the border cities grew even more quickly: between 1910 and 1940, Ciudad Juárez grew from a town with something over 10,000 inhabitants to one almost five times larger. However, when we examine these cities in 1950, the first date for which there are tenure data, there is little sign that demographic growth had transformed the dominant pattern of renting. Mexico City, for example, seems to have absorbed the vast majority of its population through rental housing: in 1950 four out of five households in the city were living as non-owners. Similarly, in Ciudad Juárez, two out of three households were living in some kind of non-ownership.[4]

Considering population growth after 1940 produces similar inconsistencies in terms of the relationship between growth rates and tenure change (Tables 3.1 and 3.3). Most cities continued to absorb very high rates of urban growth through an expansion in the rental housing stock. Although

Monterrey grew at over 6 per cent per annum during both the 1940s and the 1950s, less than one in three homes were occupied by owners in 1960. Indeed, the share of owners in Monterrey actually declined during the 1950s, a common pattern found in numerous cities. In contrast, Morelia, which grew less rapidly than Monterrey, had more than two-fifths of its homes in owner-occupation by 1960. The major shifts from renting to ownership do not match the periods of rapid urban growth in many cities. In Guadalajara, annual growth rates of around 5 per cent occurred twenty years before there was a real shift towards self-help ownership. In Torreón, huge increases in population seem not to have brought an increasing share of home-owners until the 1960s.

*Table 3.3* Annual growth rates by city, 1940–70 (per cent)

|  | 1910–40 | 1940–50 | 1950–60 | 1960–70 |
|---|---|---|---|---|
| Mexico City | 4.2 | 5.7 | 5.1 | 5.1 |
| Guadalajara | 2.4 | 4.8 | 6.8 | 5.5 |
| Monterrey | 3.0 | 6.2 | 6.6 | 5.2 |
| Puebla | 1.2 | 4.7 | 2.4 | 6.0 |
| Ciudad Juárez | 5.2 | 9.1 | 7.8 | 4.4 |
| León | 0.8 | 4.3 | 5.2 | 4.9 |
| Mexicali | 13.1 | 10.9 | 8.5 | 3.8 |
| Torreón | 3.7 | 5.0 | 2.9 | 2.4 |
| Tijuana | 10.9 | 11.5 | 9.7 | 7.5 |
| Culiacán | 1.6 | 4.7 | 3.6 | 5.6 |
| Chihuahua | 1.2 | 3.6 | 5.2 | 4.0 |
| San Luis Potosí | 0.4 | 4.7 | 2.2 | 3.3 |
| Mérida | 1.5 | 3.3 | 1.8 | 2.4 |
| Toluca | 1.1 | 1.6 | 3.1 | 4.4 |
| Acapulco | 1.8 | 9.4 | 4.2 | 10.9 |
| Veracruz | 1.3 | 3.6 | 3.6 | 4.1 |
| Aguascalientes | 2.0 | 1.3 | 2.7 | 3.8 |
| Morelia | 0.3 | 3.2 | 3.7 | 3.6 |
| Hermosillo | 0.8 | 6.1 | 8.1 | 5.8 |
| Durango | 0.2 | 4.7 | 3.8 | 3.6 |

*Sources:*  1910–40   Unikel *et al.* (1976: Table I-A1) and own calculations.
     1940–70   Scott (1982: Table 4-2), elaborated from *Censo General de Población*.

Of course, self-help housing was increasing everywhere. Given the pace of demographic growth, large areas of clandestine settlement were being created even when the proportions of renting to ownership remained the same. This can be demonstrated in the case of Mexico City where self-help housing areas were emerging from the 1940s or even earlier (COPEVI, 1977; Perló, 1981). Nevertheless, the census data indicate that, up to 1960, rental accommodation was increasing even more rapidly. Similarly, in Guadalajara, self-help housing began to emerge as a serious option for the poor during the 1940s (Sánchez, 1979). Despite the expansion of self-help ownership, however, the relative numbers of owners only began to expand in the 1960s.

The likely explanation for this diversity of experience lies in the form and dynamics of local land markets. It is obvious that different patterns of local state intervention mean that the poor were much less able to obtain land for self-help in some cities than in others. It is also likely that the local land market changed through time as a result of improvements to transport (thereby opening up new areas of land), changes in local authority attitudes to land invasion, fluctuating levels of community organisation, and increasing scarcity of land suitable for urban development (as a result of physical constraints on growth or changes in property ownership). These differences may perhaps explain both the differing tenure mixes and the timing of the move to home-ownership.

An explanation of this sort might well be adequate in the case of the northern cities such as Tijuana, Mexicali, Chihuahua, Reynosa and Monterrey. During the 1960s and 1970s, a time when land was relatively open to access by the poor, ownership became much more common in each of these cities. In Tijuana and Ciudad Juárez, for example, Hoenderdos *et al.* (1983: 385) have reported that 'abundant tracts of land are parcelled out in plots by the local authorities and issued at prices which roughly correspond to three to five months of rent for a dwelling in or near the city centre'. In addition, land invasions were common in Tijuana (ibid.: 379), and according to de la Rosa (1985: 45) 'waves of people ... invaded as "parachutists" the whole periphery of the city'. In Reynosa, there was a wide-spread growth of popular settlements, many 'the result of land invasions' (Margulis, 1981: 173). In Chihuahua, small-scale 'fill-in squatting' during the early 1960s gave rise to a much larger series of invasions after 1968 led by the Comité de Defensa Popular (Verbeek, 1987: 4). In Durango, invasions of private and *ejido* land began in the 1960s, at the initiative of PRI politicians, and became commonplace during the 1970s (Ramírez Saiz, 1986: 406).

In Monterrey, invasions were also an important source of land, even if the ease with which invasions could occur seems to have varied dramatically through time (Pozas Garza, 1989). According to Balán *et al.* (1973: 308), there had been 'relatively little illegal land seizure in Monterrey' up to 1970, although other accounts suggest that illegal settlements were being formed in Monterrey before this time. Both Ortiz Gil (1982: 165) and Villarreal and Castañeda (1986), for example, recall one case of invasion as early as 1928 and invasions were quite common during the 1960s. What is clear is that both the frequency and size of the invasions increased dramatically during the 1970s; of the invasions which have been dated, almost one-half of the residents live in settlements founded after 1970. As Ortiz Gil (1982: 57) comments: 'from 1960 the invasions accelerated throughout the metropolitan area ... the 1970s marked the beginning of the avalanche'. Between 1962 and 1967, the state more or less permitted the illegal subdivision of land without services (Villarreal and Castañeda, 1986: 36), a consequence of organisations affiliated to the PRI sponsoring 'official invasions' during the early 1960s. The 1970s, in contrast, were marked by a wave of land invasions organised by opposition groups. If the invasions had to struggle hard against repression during the early 1970s, they became easier to organise towards the end of the decade (Pozas Garza, 1989). In response to the wave of invasions the state introduced a pro-gramme to control the situation. While this was less than successful in terms of extracting land payments from the poor, by 1979 it had settled many more families on their own plots (Ortiz Gil, 1982: 127). It is surely not coincidental that Table 3.1 shows a major increase in the proportion of home-owners in Monterrey during the 1970s.

In certain other cities of Mexico, however, land invasions have been infrequent and other mechanisms have developed to accommodate the poor. In Mexico City, 'invasions are of limited importance as a method of land alienation' (Gilbert and Ward, 1985: 91). In the Federal District, Governor Uruchurtu resisted attempts to invade land between 1953 and 1966 and even in the State of Mexico, the normal process of land allocation for the poor was through illegal subdivision. Access to land for the poor has been firmly controlled in the State of Mexico since 1973 and in the Federal District since 1977 (ibid.: 96).

In Guadalajara and Puebla there has also been little in the way of land invasion (Logan, 1979: 133; Castillo, 1986; Mele, 1986b: 43). In the absence of free land, the principal alternatives have been the illegal sub-division of private land and the illegal purchase of *ejido* land (see Chapter Five). Both mechanisms have developed in many of the larger Mexican cities. In Mexico City, illegal subdivisions were the principal form of land occupation to the east of the city, at least until the 1970s (Gilbert and Ward, 1985). Municipal authorities in the State of Mexico connived with the land developers to encourage the conversion of what previously had been national property. Bought cheaply during the 1920s for agricultural development, the land was ripe for subdivision and sale to the poor. In addition, between 1940 and 1982, 27 per cent of the city's expansion took place on *ejido* lands (Varley, 1985a: 3). In certain parts of the city, over half the population was resident on *ejido* lands (Varley, 1989a: 132).

In other cities, too, *ejidos* have been a principal source of urban land. In Querétaro, the massive occupation of *ejido* land became common after 1969 (García Peralta, 1986: 380). Similarly, in the vicinity of the oil towns of Veracruz and Chiapas, *ejido* land has been subdivided, sometimes with the participation of the local authorities (Legorreta, 1983: 66). While there have been a few cases of invasion, the local authorities have usually permitted illegal development both for reasons of political control and for the illicit gains to be made from involvement in the subdivision process. Although the general format seems closer to that of the illegal subdivision than of the invasion, the prices charged for the land have been rather low (Legorreta, 1983: 75, 77).

The fact that land invasions have been relatively rare in Mexico City, Guadalajara, Puebla and Querétaro has led to different forms of land alienation emerging. The significance of that development for the poor is that it makes land acquisition more expensive. Perhaps this fact alone could explain why all such cities retained high levels of non-ownership in 1970 and 1980 (Table 3.1). An irrefutable link is difficult to establish because land allocation processes are very complex and because it is not always easy to distinguish between different kinds of illegal land occupation. Nevertheless, there are clear differences between the land markets in different Mexican cities, and the possibility of a link between the cost of land and the incidence of ownership will be examined below.

## Explanations of the incidence of land invasions

It is possible that the differences between local land markets and the reactions of the state to land invasions may be linked in some way to local physical and climatic conditions. Is it wholly coincidental that many of the cities which have been subject to frequent land invasions are located in dry areas? Indeed, if we rank cities by the amount of rain they receive, then a large number of the cities with high levels of home-ownership are located in the drier areas. In Chihuahua, Verbeek (1987: 4) has attributed the ease with which *ejido* land was occupied to the fact that these 'communal agricultural grounds consisted mainly of harsh dry lands, unsuitable for any kind of cultivation, and of little value'. It is equally clear that certain cities with high ownership are located in very wet areas – for example, Mérida, although not Veracruz. The common element is that these cities too are surrounded by land of low agricultural potential; the authorities are far less likely to defend desert or forested land than land of high agricultural value.

A further explanation may be linked to the urban transport situation. Clearly, it is difficult to live in peripheral settlements if there is no form of public transport linking the settlement to zones of employment. In many cases, invasions take place on land which is distant from the city centre. While it is difficult to evaluate the quality of public transport or make general statements about the journey-to-work situation in different cities, there is one feature of the border cities which is likely to encourage the development of peripheral settlement (if not necessarily land invasions). The available data make it clear that the border states, and particularly Baja California, have very high levels of private car ownership (Herzog, 1989: 121). This is a direct outcome of higher incomes in the border cities and of the availability of cheap, used cars from the USA. If one of the factors that impeded the growth of self-help settlements in most developed countries in the nineteenth century was the lack of urban transportation (Stedman-Jones, 1971), this is not a constraint on low-density suburbanisation in the Mexican border areas.

Finally, there may be a relationship between the level of political protest by the poor and the level of state support for, or tolerance of, illegal land alienation. In the history of rent strikes some cities seem to be more combative than others. Similarly, social movements appear in some cities more frequently than others. It is plausible that where left-wing political groups become very well organised, as in Chihuahua or Monterrey during the 1970s, the incidence of land invasions or state provision of cheap plots may increase. The corollary is also likely: illegal land development processes are likely to be less common in cities which are widely considered to be dominated by conservative elites and ideology. Both Guadalajara and Puebla fall into this category and neither has experienced land invasions. Is the lack of illegal subdivisions and particularly the lack of land invasions due in some part to the role of the Church and the conservative ideologies of those cities? Clearly, there have been many efforts to mobilise the poor in both cities (Castillo, 1986; Regalado, 1987), although the effectiveness of this organisation has been less successful than elsewhere. Certainly, associations from these two cities have participated little in confederations of social movements such as the CONAMUP (Ramírez Saiz, 1986).

In sum, it is difficult to explain concisely why there are such wide differences in the incidence of land invasion in different parts of Mexico. Clearly, much of the answer lies in local differences in political economy. Local political conditions, the opportunity cost of land, the pattern of land ownership, the rate of population growth, the local investment climate, the transport situation, and the level of social organisation all explain part of the difference. Nevertheless, the impact of that difference on the poor is not unimportant in so far as it affects the cost of land. While some of the poor may not want to bear the burden of self-help ownership, the likely take-up of that option is bound to be affected by the cost of peripheral plots relative to local rent levels. It is our contention that the price of land is a key ingredient explaining differences in patterns of residential tenure between cities.

## CONCLUSION

Local housing markets are an important element influencing living standards. Current British experience also suggests that local housing differences may affect labour mobility and hence national economic performance. It is therefore important to understand the working of these markets. A critical dimension of the local housing situation is residential tenure. In Mexico, there was a major national shift from rental tenure to ownership in the period from 1940 until 1980. And yet, there were important variations in the local pace of change. In 1980, half of all homes in Puebla were rented or shared compared to one-fifth of all homes in Culiacán. Since most Mexicans say they aspire to home-ownership, this is a significant difference in local housing conditions.

This chapter has attempted to shed light on some of the factors influencing local variations in residential tenure. It is by no means a definitive account, for work on this topic is ongoing. Nevertheless, it is already clear that there are important links between housing tenure and the wider political economy. Residential tenure patterns in Monterrey or in Mexico City cannot be understood without a detailed understanding of local political and economic realities. Such a statement is hardly novel, but nevertheless it is still too common for experts on housing to neglect the political and economic context when making their policy recommendations.

# 4    Mexican housing policy

The dramatic shift from renting to ownership in urban Mexico was, and continues to be, strongly encouraged by the state. State policy has stimulated the growth of home-ownership, both directly and indirectly, while neglecting, and sometimes even discouraging, most forms of rental housing. In the formal sector, public agencies have provided funds mainly for the construction of owner-occupied homes; in the informal sector, the principal form of state action has been the gradual improvement, regularisation and servicing of irregular settlements.

The state has done little to stimulate the construction of rental housing and has failed to do much even to maintain the existing stock. Since 1963, no public agency has constructed housing for rent and the small stock of state rental housing has recently been offered for sale. In the private rental sector, the state's main influence has been through legislation. Unfortunately, the combination of rent controls, building regulations and financial incentives has discouraged private landlords from investing in this form of housing.

The purpose of this chapter is to summarise the main elements of Mexican housing policy, particularly as it relates to rental accommodation. The first section examines public housing finance; the second, state policy towards informal housing; the third, state attitudes to building housing for rent; the fourth, the legislative framework governing the rental sector; and the final section, the government's efforts to improve rental housing in the central city.

## PUBLIC HOUSING CONSTRUCTION

Since the government first began to build homes in 1925, its policy towards home construction can be divided into five distinct phases. The beginning of each phase is normally associated with the establishment of new housing agencies intended to rectify past errors or to expand the housing programme. As Figure 4.1 shows, however, existing agencies have rarely been abolished: they generally have been allowed to carry on during the new phase, admittedly with smaller budgets.

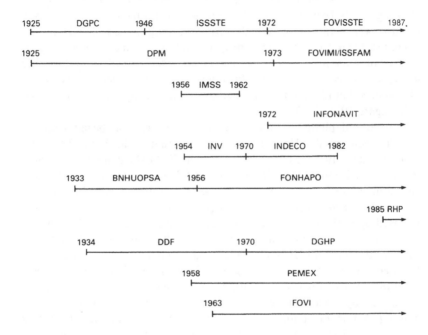

*Figure 4.1* Public housing agencies, 1925–87

*Source*: Adapted and updated from Garza and Schteingart (1978: 80).

The first policy phase ran from 1925 until the mid-1950s. It consisted of the establishment of a number of agencies responsible for building housing for favoured government workers. The General Directorate for Civil Pensions (DGPC) was established to build housing for federal government workers and the Directorate of Military Pensions (DPM) to cater for the military. Later, other public sector employees obtained their own agencies, notably those working for the Federal District, the petroleum sector, and the sugar industry; eventually all government workers became eligible for subsidised public housing. Although the coverage of public employees expanded quite quickly, in practice, few workers received a home because none of the agencies constructed enough houses (Table 4.1). Admittedly, some effort was made to increase the pace of housing construction with the establishment of the National Mortgage Bank for Public Works in Urban Areas (BNOPSA) in 1933, but this agency did relatively little until it was reorganised as the National Bank for Public Works (BANOBRAS) in 1947. Its structure was further modified in 1954 when the Trust for Popular Housing (FONHAPO) was established within the bank. This Trust was given power to raise money both through issuing housing bonds and by obtaining loans from private banks, methods of finance which were to become of major importance in succeeding years. Also in 1954, yet another institution, the National Housing Institute (INVI), was added to the plethora of agencies; but like the rest, it failed to produce many homes (Garza and Schteingart, 1978: 131).

The second phase in Mexican housing construction began with the establishment of the Housing Finance Programme in 1963. This programme was distinctive in so far as it was the first to be generously funded. It obtained substantial sums from the United States government, which was channelling money into Latin America as part of the Alliance for Progress programme. It also benefited from a programme of compulsory lending imposed upon the public banks, the latter being required to place 30 per cent of their total savings into two trusts established within the Bank of Mexico: the Fund for Bank Operation and Discount (FOVI) and the Fund for Guarantee and Support for Credit for Housing (FOGA). These moneys were intended to finance the building of homes for lower- and middle-income groups, accommodation that became known as 'social-interest' housing. While the policy was very successful in stimulating housing construction (Table 4.1), it did little to improve access to housing since few workers could afford to buy the housing being provided. By the late 1960s, when it was becoming clear that housing conditions were deteriorating in the major cities, the Mexican Workers' Confederation (CTM) began to pressure the government to change its policy (Ward, 1988).

The third policy phase was an outcome of this pressure. The incoming administration of Luis Echeverría began to give strong support to the regularisation and improvement of self-help housing (see next section), and sought to increase the production of 'social-interest housing'. The principal innovation in terms of housing finance was that pension funds were to be tapped for the first time. Three major institutions were established for this purpose: the National Institute for Funding Workers' Housing (INFONAVIT) to build housing for private sector employees; FOVISSSTE for federal government workers; and FOVIMI for the military. This giant injection of funds allowed the Mexican government to increase the construction of 'social-interest' housing at relatively little additional cost to the public purse. This also cut the level of dependence on the private banks, bank credit falling from 85 per cent to about half of the total housing investment (Garza and Schteingart, 1978: 228).

*Table 4.1* Public housing finance: units constructed, 1947–88

|  | 1947–64 | 1965–70 | 1971–79 | 1980–86 | 1983–88 |
|---|---|---|---|---|---|
| *Social security agencies* | | | | | |
| ISSSTE | 45,302 | 1,300 | 4,375 | – | – |
| IMSS | 10,600 | – | – | – | – |
| Others | 1,100 | n.d. | 764 | – | – |
| *Housing funds* | | | | | |
| INFONAVIT | | | 213,785 | 378,700 | 414,206 |
| FOVISSSTE | | | 60,811 | 55,723 | 92,658 |
| FOVIMI/ISSSFAM | | | 1,965 | 3,222 | 2,549 |
| *Public housing agencies* | | | | | |
| INVI/INDECO | 10,600 | 3,800 | 62,695 | – | – |
| BANOBRAS/FONHAPO | 24,098 | 16,644 | 19,550 | 192,719 | 245,068 |
| FOVI (direct construction) | | 15,572 | 5,956 | – | – |
| Earthquake relief | | | | 52,590 | 66,165 |
| *Others* | | | | | |
| PEMEX and DDF | 29,500 | 6,000 | 34,454 | 47,312 | 105,908 |
| *Private housing* | | | | | |
| FOVI/FOGA financing | | 76,443 | 106,689 | 279,491 | 468,637 |
| Total | 121,200 | 119,759 | 511,044 | 1,009,757 | 1,395,191 (903,287) |
| Annual total | 6,733 | 19,960 | 56,783 | 144,251 | 232,532 (150,548) |

*Sources:* Garza and Schteingart (1978: 80), Ward (1988: Tables 2 and 3), Mexico, SEDUE (1989a: Tables 1 and 2) and Duhau (1988: 37).

*Note:* There is a considerable amount of double and even treble counting in the figures for some agencies (particularly INFONAVIT and FONHAPO), caused by uncompleted buildings being carried over from one year to the next. The figures for 1980–86 and 1983–88 are impossible to separate meaningfully, so both have been included. The figures for 1980–86 and 1983–88 include all schemes including sites-and-services projects, credits and improvements. The bracketed final total figure for 1983–88, however, only records whole units, thereby excluding such schemes. Clearly, the figures after 1980, therefore, are not strictly comparable to those for earlier periods.

The fourth policy phase began as a result of the onset of the recession in 1982. Faced by a massive fall in government revenues, the administration of Miguel de la Madrid responded by cutting subsidies in the housing sector. Gradually, the government has been raising the charges to borrowers so that they now pay the full cost of construction and credit. Rather than cutting the rate of construction, however, the government decided to favour the housing sector as a way of maintaining levels of employment. Faced by rising levels of urban unemployment, the National Programme for Housing Development sought to mobilise funds for construction equivalent to 1 per cent of the gross domestic product. The banks, which had been nationalised in 1982, were forced to invest 3 per cent of their reserves in housing and, as a result, more than 900,000 houses were built between 1983 and 1988 (Mexico, SEDUE, 1989a: Table 2).

The problem during the 1980s, therefore, has been less the pace of construction than the inability of the mass of the population to pay for the new housing. With inflation increasing dramatically and the government holding back rises in the minimum salary, the purchasing power of most wage-earners has been cut severely. Between 1980 and 1989 the purchasing power of the minimum salary fell by 49 per cent (UNECLA, 1989: 18). Household incomes among middle-class households probably fell even faster (Mexico, INC, 1989). When combined with rises in the real price of land and materials, this means that the real cost of 'social-interest' housing has risen dramatically: there are currently few households which can afford to pay off the vastly increased cost of a loan. In 1978, for example, 'social-interest' housing in the three major cities was accessible to those earning between 5.0 and 5.6 times the minimum salary; by 1984, that same housing required an income somewhere between 18.2 and 22.1 times the minimum salary (Mexico, FOVI, 1986: 61–2). The position of lower-income households has been made worse by the decision to phase out subsidies on some kinds of 'social-interest' housing. Since 1984, for example, FOVI has started to charge borrowers the full cost of attracting funds (Mexico, FOVI, 1986: 47; Mexico, FOVI, 1989: 25).

The final phase in housing policy overlapped partially with the fourth and was a direct outcome of the consequences of the 1985 earthquakes. Faced by terrible devastation in the centre of the capital city and by the massive mobilisation of those affected, the government responded in a highly novel and effective manner (Azuela, 1987; Connolly, 1987; Duhau, 1987; Massolo, 1986). It expropriated all the rental property that had been badly damaged by the earthquake, provided temporary accommodation for the inhabitants and established a programme to reconstruct housing in the central area. President de la Madrid signed the expropriation decree in October 1985 after several groups of 'refugees' had marched to his residence demanding action. As the Journal of Commerce reported: 'the measure was taken in response to the unusual mass mobilisation of poor downtown residents who feared eviction from the condemned property' (*Los Angeles Times*, 16 October 1985).[1] The decree offered compensation, payable over a period of ten years, and expropriated more than 5,000 properties occupying 625 acres of the central city (Azuela, 1987). The expropriation was highly unusual, as was the reconstruction of housing in the central city; before the earthquake the idea of the federal government subsidising the cost of rebuilding housing in the central area would have been inconceivable. Unprecedented or not, within four years around 70,000 new homes had been constructed, some 44,500 by Popular Housing Renovation (RHP, a subsidiary agency of FONHAPO), and the rest by the Tlatelolco Reconstruction and Phase II Emergency Housing programmes (Michel, 1988: 14). This temporary departure from normal practice was highly successful, being praised internationally, by Mexican commentators on both the left and right of the political spectrum, and by the new homeowners. With its achievement widely lauded, Popular Housing Renovation was wound up on schedule in 1988. The abolition of this highly effective agency underlined the fact that the Mexican government did not want to extend the programme to areas of deteriorating housing which had been unaffected by the earthquake. Although a second phase was launched in 1987, this could not accommodate all those who were living in housing that had been falling down slowly for years. Nor was the new policy extended to other Mexican cities; indeed, officials in the latter feel that they have been deprived of funds due to the reconstruction going on in the capital.

## GOVERNMENT POLICIES TOWARDS SELF-HELP HOUSING

For many years, self-help housing developed in a largely unregulated environment. Although, in principle, the government was hostile to self-help housing because it was not properly serviced and often involved the illegal use of *ejido* land or the outright invasion of private property, in practice the authorities often responded sympathetically (Cornelius, 1975; Gilbert and Ward, 1985; Varley, 1985b). When land was subdivided illegally, the authorities mostly looked the other way. While there were exceptions, as in the Federal District between 1952 and 1966, self-help settlements were allowed to develop in most Mexican cities. Gradually, government agencies agreed to provide services and infrastructure. Political pragmatism recognised that however irregular the self-help process was, to stop it would make the housing and political situations much worse. Benign tolerance was therefore perceived to be the most sensible response.

In 1970, this approach became explicit when the Echeverría administration established a series of new agencies to help regularise self-help housing settlements (Ward, 1986; Varley, 1985b). INDECO and CORETT were established at the federal level and other agencies were set up to operate at local level. This new approach was prompted in part by a crisis in the Ciudad Netzahualcóyotl area in the State of Mexico, where the lack of services and infrastructure had prompted local residents' associations to mobilise and protest against the actions of the land developers (Ward, 1986). Any threat of settlement demolition seemed to be removed as this gamut of agencies went into action. While this action was highly politicised, the point was that the state was seen to be explicitly supporting self-help settlement rather than opposing it. Not surprisingly, some interpreted this approach to mean that the process of illegal land occupation could continue, and as a result more land was urbanised and further invasions took place.

Echeverría's populist policies gradually gave way to a more restrictive strategy. Since the late 1970s, the federal government's policy has been to control and direct the process of irregular settlement. The main aim has been to slow the occupation of peripheral land and to intensify land use within the urban area, thereby increasing population densities and easing the task of service provision. Associated with this approach has been a programme of land regularisation, the issuing of title deeds, and the gradual introduction of infrastructure and services. In addition, FONHAPO has been trying to accelerate the process of organised self-help by offering credit to organisations prepared to sponsor it; something like 110,000 progressive housing solutions and 75,000 serviced sites were reported to have been provided by this agency between 1983 and 1988 (Duhau, 1988: 37; Mexico, SEDUE, 1989a: Table 1). The broad aim, therefore, has been to legalise and service existing housing and to slow the creation of new illegal settlements by creating large new areas of progressive housing development.

Since 1983, the Ministry of Urban Development has also been trying to create land reserves around the periphery of the larger cities, in an effort to control the conversion of *ejido* land and to slow the outward expansion of Mexican cities (Rébora, 1986). By the end of 1986, reserves of 2,659 hectares had been established around nine major cities (Mexico, SEDUE, 1987). Of course, the question is whether this policy can ever be fully implemented because of the inherent conflict that it involves between urban and 'rural' interest groups. In so far as the mediation of land disputes has long been used by politicians as a means of winning electoral support (Cornelius, 1975; Gilbert and Ward, 1985; Ward, 1986), the new policy towards low-income settlement requires a gradual shift from the populist approach of 'politicians' to the more managerial approach favoured by 'technicians'. And because the Salinas government is firmly committed to such a shift in government management generally, the policy may soon become more effective. Certainly, the occupants of a large illegal settlement in Mexico City, Lomas del Seminario, were removed by the previous administration in November 1988.[2] If this policy becomes fully effective in the future, then the implication is that home ownership is likely to become more expensive for the poor. If legal, semi-serviced plots constitute the only source of new land, the cost of access will undoubtedly rise. In turn, this is likely to increase the propensity to rent and share accommodation.

On the other hand, there are three major reasons to suppose that this general approach will be less than successful in controlling the growth of illegal settlement. First, the Mexican government continues to encourage the growth of home-ownership. And, while the constitution decrees that every home should be 'decent and clean' (*digna y decorosa*), political realities will probably dictate that any home is better than no home. Second, the current political environment is one where the PRI is under much greater electoral threat than ever before (Cornelius *et al.*, 1989). Political realities may therefore dictate that land be managed pragmatically rather than technically. Third, the whole process of land and settlement management presupposes that the state is able to deliver infrastructure and services. While such service delivery improved markedly during the 1960s and 1970s, the recession is posing increasing questions about servicing in the future. Given the debts of the Federal District administration, for example, it is difficult to see how capacity can be increased. And, since current policy is hostile to subsidies, any expansion in service provision will have to recoup the cost from consumers.

The future, therefore, is likely to be rather like the recent past. Despite the understandable desire to regulate and direct the process of self-help settlement, the combination of political needs and financial constraints are likely to favour the adoption of a more 'realistic' approach. A mixture of permissiveness and repression will lead to the continued incursion of urban settlement onto *ejido* land, interspersed occasionally by flurries of government repression. Regularisation policies will continue, but the rate of servicing will slow.

## THE BUILDING OF RENTAL HOUSING

In contrast to the huge effort that has been made to finance the construction of houses for sale and, more recently, to encourage self-help housing, the state has failed to stimulate the rental sector. Indeed, many have argued that

state intervention, particularly through rent-control and planning regulations, has hastened the decline of this kind of housing. In addition, the
Mexican government has latterly refused to build public housing for rent.
Indeed, discussion of rented housing was for several years excluded from
the policy agenda. It was not until the López Portillo *sexenio* (1976–82) that
the state once again began to discuss the need for more rental housing, only
under de la Madrid that it began to find ways of encouraging private rental
construction.

The history of state housing for rent is short and undistinguished. In
1935, BNHUOPSA proposed that the state should finance rental housing
for workers (Coulomb, 1985a: 43). It was not until 1949, however, that the
*Dirección de Pensiones Civiles* and its successor institution, ISSSTE,
began to construct apartments for rent. A similar policy was taken up soon
after by IMSS. Between them, the two institutions built some 18,000 units,
the vast majority in Mexico City, during the 1950s. In other parts of the
country (for example in Guadalajara) state governments also built some
rental housing, but on nothing like the same scale (González Navarro,
1974).[3] Unfortunately, the experience proved to be highly problematic for
the institutions concerned: the agencies found it difficult to raise rents and
were soon confronted by major losses on their housing programmes (Garza
and Schteingart, 1978). By 1963, the programmes were suppressed and
both IMSS and ISSSTE restructured; the decision was taken not to let
accommodation in the future. Indeed, a resolve that the state should never
again become a landlord seems to have underlain public housing policy
ever since.

It is only during the past ten years that rental housing has regained a
place on the political agenda, a change attributable to the rising cost of
home-ownership in the major cities. With many among the working and
middle classes beginning to reconcile themselves to being tenants, the
shortage of adequate rental accommodation became a major source of
complaint. Given the general consensus in the construction and real-estate
sectors that 'renting is about to become history because it is no longer good
business', investors had simply stopped putting money into rental housing
(*El Financiero*, 2 December 1983). As a result, the state was forced to take
some kind of action.

In 1978, after some years of silence, the National Housing Programme
actually included a reference to rental housing. It announced that the
existing rental stock should be maintained and recommended that
encouragement be given to private investors. The Human Settlements
Ministry began to consider different kinds of fiscal incentive for builders,
particularly in the field of 'social-interest' housing (Mexico, SAHOP,
1978: 157). In August 1980, the availability of tax-relief certificates

(CEPROFIS) was announced for companies producing 'social-interest' housing. Although the incentives appeared to be particularly generous for housing intended for rent, it soon became clear that little new investment was being attracted.[4] An internal SEDUE paper noted that between 1977 and 1982 only eight buildings had been constructed for rent in the Federal District (*El Financiero*, 2 December 1983). As a result, the scheme was abolished at the end of 1982 and replaced, at the end of 1984, by seemingly more attractive incentives. FOVI began to offer loans for up to 70 per cent of the cost of housing construction at a rate of interest no higher than 14 per cent per annum. The main conditions on the loans were that the accommodation be rented for a minimum period of ten years and that the rent should not exceed the official minimum salary (Mexico, FOVI, 1986: 18–19).[5] FONHAPO also introduced similar kinds of incentives for rental housing (Mexico, FONHAPO, n.d.: 38). In addition, further encouragement was given in the form of accelerated depreciation allowances. These incentives were made still more attractive in September 1985 when CEPROFIS were extended to companies building housing that would be rented for a minimum of five years; tax relief equivalent to 15 per cent of the total value of the investment would be available (Mexico, SEDUE, 1987).

Table 4.2 suggests that the initiatives have been effective. Unfortunately, the table contains figures for permissions to build rather than data on actual construction. Most sources suggest that the difference between the two is rather large, of the order of five to one. This interpretation is supported by the most recent figures produced by FOVI and SEDUE. Two separate sources suggest that only 60,000 rental homes were begun during the 1982–88 *sexenio* of which 43,641 were completed (Mexico, FOVI, 1989; Mexico, SEDUE, 1989a: Table 2). The housing that has been completed has been concentrated mainly in rapidly expanding tourist centres such as Cancún and Puerto Vallarta. While there has been some building for rent in Guadalajara (Mexico, FOVI, 1986), there has been little activity in Puebla or in Mexico City. In most cities, investors seem to have preferred to put their money into other kinds of activity. Certainly, recent government comments about the success of the scheme have been rather equivocal. During 1987, FOVI and SEDUE claimed that the programme was proving a great success but also recognised that it would operate more successfully if the administrative system were reformed. 'In spite of the fiscal and credit incentives that have been established to stimulate rental housing, there is still a series of administrative barriers that are discouraging investment by small investors and by the private sector in general' (Mexico, SEDUE, 1987). Recently, the Ministry admitted that, in spite of the incentives, 'it has not been possible to achieve the anticipated levels of investment'.

Nevertheless, the Ministry continues to have frequent discussions with the building companies about further incentives in the hope of stimulating further investment.

*Table 4.2* VIS-R credits approved by FOVI, 1984–87

|  | 1984 | 1985 | 1986 | 1987 |
|---|---|---|---|---|
| Mexico City | 1,749 | 4,902 | 12,562 | 4 |
| Guadalajara | 2,490 | 4,074 | 7,482 | 64 |
| Monterrey | 4 | 2,108 | 6,715 | 102 |
| Intermediate cities | 5,231 | 16,275 | 49,904 | 5,235 |
| Others | 1,092 | 565 | 1,155 | 0 |
| Total | 10,566 | 27,924 | 77,818 | 5,405 |

*Source:* FOVI unpublished data.

*Note:*   Figures for 1987 refer only to January to May.

Recent efforts have also been made in the Federal District to stimulate rental housing. The reform of rental legislation issued in February 1985 (see below) decreed that the local authority, the DDF, encourage the development of rental housing. Unfortunately, as Coulomb (1985a: 36) points out, the DDF does not have the resources to do this; any action can only be taken through FOVI and FONHAPO. The decree would seem, therefore, to be more a sign of wishful thinking than of effective policy.

The general lesson seems to be that it is currently difficult to attract Mexican capital into rental housing. This situation is likely to persist so long as investors perceive that the legislation continues to favour tenants. Even more important is the availability of more attractive sources of profit. Certainly, recent years have provided an abundance of such alternatives: shares, dollars and even bank savings. Between December 1982 and October 1987, for example, the average real price of shares on the Mexico Stock Exchange almost doubled. Even after the effects of the great October crash, investors would still have made an 80 per cent gain over the whole de la Madrid *sexenio* (*El Mercado de Valores*, 1 December 1988). In addition, bank interest rates have offered a real return for long periods, an attractive alternative at a time when business prospects seemed uncertain. Finally, when share prices and/or interest rates were falling, the attractions of capital flight remained an ever real option.

## HISTORY OF RENTAL LEGISLATION

Official controls over the level of rents, sanitary conditions, rental contracts and so on have a long history in Mexico. As early as 1870, the Civil Code for the Federal District laid down rules on the length of tenure, and required that a written contract be signed when the rent exceeded a certain level (Parcero López, 1983). During the Mexican Revolution, tenant protests encouraged different 'national' governments to issue rental decrees (Perló, 1979: 772–4). It is clear that the complaints of tenants have long been a source of political problems for the authorities (González Navarro, 1974: 174–5). Such complaints have always been most vociferous during periods of recession or rapid inflation; at such times, politicians have sometimes been forced to take action, whether it be effective or otherwise. Given the frequency of landlord-tenant disputes and the likelihood that these might spill over into tenant strikes such as those in Mérida and Veracruz in 1922, Mazatlán in 1935, and Tepic in 1938, it is perhaps surprising that there is no national legislation on rental housing. Although an early effort was made to pass a Federal Rent Law in 1921, this was unsuccessful (*Proceso*, 19 September 1983). Perhaps Perló (1979: 774) is correct when he explains this failure in terms of the tenants' lack of political weight; there may have been large numbers of tenants but their votes could always be won in other ways. An alternative explanation is that the lack of a federal law arises from the wording of the National Constitution: legislation on renting, it declares, is the responsibility of the states. This enshrinement of local autonomy is intended to guarantee that rental legislation confirms local conditions (Parcero López, 1983: 106).

The form of official control over renting, therefore, has always varied from state to state. Following the Revolution, Mérida introduced rent controls in 1922 and Veracruz instituted a rent freeze in 1923 (Perló, 1979: 778–9); in contrast, the authorities in the Federal District, despite the submission of numerous proposals, failed to approve any legislation. The effects of the recession of the 1930s brought a legislative response in Hidalgo, Tamaulipas and Veracruz, but not elsewhere.

While local variations are still important, the extent of inter-state variation should not be exaggerated. Increasingly, a tendency has been apparent for the provincial states to follow general practice in the Federal District. Such was certainly the case with arguably the most influential pieces of Mexican legislation, the rent freezes of the 1940s. Introduced first in the Federal District in 1942, the example was followed by a number of local legislatures: Puebla in 1943, and Tabasco and Yucatán in 1947 (González Navarro, 1974: 193). The 1940s rent freezes are still in operation to this day. Even if they now affect relatively few properties, the existence of the legislation has long helped to sustain a feeling among landlords that the state is unsupportive of their interests. For this reason, it is worth considering the legislation in some detail.

The first rent freeze was introduced in Mexico City when the country entered the Second World War in 1942; similar rent freezes were being introduced elsewhere in Latin America around the same time.[6] It was introduced at a time when the Mexican economy was suffering from severe inflation, the decree being intended as a temporary remedy for falling living standards (Bortz, 1984).[7] The original decree was extended on several occasions and continues today on the basis of a final extension in 1948. COPEVI (1977: 25) have argued that the rent freeze was extended as part of a general policy aimed at holding down industrial wages: strong union pressure could only be assuaged by such a measure. This is probably too simple an explanation because it fails to explain why both the original freeze and later extensions to the freeze included only certain groups and particular residential areas of the city. Perló (1979) explains this limited coverage in terms of the political interests of the PRI. While the rent freeze was intended to win political support among tenants, there was no wish to alienate the majority of urban property owners. Unlike the rest of the Federal District, the central area contained large numbers of beneficiaries and relatively few landlords. Even more advantageous was the fact that among the central tenants were many small traders and manufacturers. Any loss of landlord support would be more than compensated by the votes of this group. While extending the rent freeze to the whole urban area would please the tenants, it might cost the PRI the support of an important group of property owners.

Many writers have argued that the rent freeze was an important influence in the decline of rental housing in Mexico City (Aaron, 1966; Grimes, 1976). However, if the rent freeze,

> contributed to a deterioration of living conditions in rental housing in the Federal District ... the negative impact of this decree tends to be exaggerated. It is blamed for the disinterest of the private sector in the production of rental housing, when the real reasons for this retraction of investment ought to be sought in the general financial and economic environment.
>
> (Coulomb, 1985a: 11)

Perhaps the major effect of the freeze was to slow the pace of land-use change in the centre of the Federal District, mainly because it complicated and slowed the transfer of low-cost housing into more profitable uses (Aaron, 1966; Connolly, 1982). As we have already suggested, however, it had little effect on most of the city. By 1961, it affected only 22 per cent of rental accommodation and 13 per cent of all housing in the Federal District (Mexico, SHCP, 1964); by 1976 it is estimated that it affected less than 1 per cent of all homes in Mexico City (COPEVI, 1977: 30).[8]

Legislation on rents was also introduced in other parts of the country. In 1943, the State of Puebla introduced an extraordinary tax to be levied on any landlord increasing the rent, the amount charged being equal to the rise plus 10 per cent. This rent freeze was slightly relaxed six months later, when rises of up to 2 per cent were permitted. In Jalisco, certain rents were frozen in October 1942 and attempts were certainly being made in 1944 to maintain this freeze. It was not long, however, before the governor was complaining of the 'excessive rents' being charged by landlords (García Barragán, 1947: 82). In general, rent control legislation had more impact in Mexico City than elsewhere. Indeed, by 1961 the effect in other parts of the country was minimal. For example, only 1.0 cent of the rental housing in Guadalajara, 4.3 cent in Monterrey, 4.8 cent in Tijuana, and 0.9 cent in Ciudad Juárez was affected (Mexico, SHCP, 1964).

## LEGISLATION DURING THE 1980s

After the flurry of rent-control activity in the late 1940s and the limited action taken by a number of states in the 1950s, very little further legislation was passed until the 1980s. In the Federal District, no new legislation was approved between 1948 and 1985, although there had been no shortage of proposals (Coulomb, 1985a: 44). In 1984, a tenants' representative claimed that: 'In the last ten years, at least 27 proposals for tenant laws have been frozen by the legislatures, which claimed that the proposals lacked the necessary judicial "elements"' (*Proceso*, 9 January 1984).

Towards the end of 1983 a joint Commission of the Senate and of the House of Deputies began working on the rental issue and ten different bills were introduced by different political groups (*Proceso*, 9 January 1984). This renewed burst of activity was stimulated by rising inflation. During 1982 retail prices rose by almost 100 per cent and tenant organisations began to complain about excessive rent rises. The National Confederation of Tenants and Settlers (CNIC) demanded a rent freeze in February 1983 and the head of the Joint Senate/House of Deputies Commission claimed that the rent situation was out of control: 'The landlords don't respect the Civil Code, which establishes that annual rent increases should not be greater than ten per cent; in practice, there are cases where rents have risen by 300 per cent' (*Proceso*, 19 September 1983).

The joint Commission was an attempt on the part of a group of PRI deputies to win electoral support. Their proposals were strongly pro-tenant: for example, three-year contracts terminable by the tenant within two. months, the banning of guarantors and deposits, and a prohibition on rental accommodation being converted into condominiums.[9] Similarly, the rhetoric supporting the proposals was strongly anti-landlord, listing in detail the many tactics used by landlords against tenants. The proposed legislation required that the DDF uphold tenants' rights. It also recommended the establishment of a free legal advice service and a system of fair rents.

In the event, the proposed legislation was not even submitted to the House because the majority of PRI legislators refused to back it (*Proceso*, 5 December 1983). Instead, they supported a presidential proposal for a Federal Housing Law, despite the fact that 'it contains nothing explicitly to protect the tenant' (*Proceso*, 9 January 1984). The head of the Joint Commission, José Parcero López, defended the new law on the grounds that it would open the way to a later initiative in defence of tenants. In contrast, the tenants' organisations were highly critical. CNIC argued that the new law 'isn't going to solve the problem ... the creation of "social-interest" housing will benefit only a small group of workers' (*Proceso*, 9 January 1984).

The situation from the tenants' perspective nearly became far worse. In December 1983, modifications to the Federal District's Civil Code were halted at the last minute. A tenants' representative claimed that the modifications would have led to some 200,000 tenant families being evicted. 'More as a result of neighbourhood pressure than out of their own accord, the deputies were forced to rectify the legislation' (*Unión de Colonos de la Colonia Guerrero*, quoted in *Proceso*, 17 February 1984).

As rapid inflation continued, rising rents stimulated further tenant demands for modifications to the Federal District's Civil Code. In February 1985, a new decree limited rent rises to the equivalent of 85 per cent of the rise in the minimum salary. It also gave the tenant a guarantee of one year's occupancy renewable for two more years and the right to take out a new contract. Tenants and landlords were henceforth obliged to comply with the stipulations laid down in the Law of Protection for the Consumer (*Proceso*, 22 December 1985). In addition, the Federal Attorney for the Consumer was henceforth instructed to represent, look after and advise the tenants. The Attorney is now required to give free advice, receive complaints, act as conciliator between landlords and tenants, and encourage the development of tenant associations (Coulomb, 1985b: 8–9). The decree also created a new kind of magistrate concerned only with rental property (the *Juzgado del Arrendamiento Inmobiliario*).

In theory, the 1985 decree should have helped tenants. As is so often the case in Mexico, however, there is disagreement over the precise impact of the reform. For a start, there seems to have been considerable confusion about the wording of the controls on rent rises. In fact, this seems to have worked to the tenants' advantage; the tenant organisations having convinced the Federal Attorney that only modifications made to the minimum salary in January should be used to calculate the rent rises. Since in recent years the minimum salary has often been modified twice annually, this interpretation means that rents can be raised much less than the total annual increase. Not surprisingly this has angered the landlords, especially since the decree seems to be quite clear as to its intention – any rise in salary should allow an increase in rent.

In other respects, however, the changes have been less clearly favourable to the tenants. Landlords continue to take tenants to court where 'judgements are generally lost by the tenants' (Lic. Manuel Fuentes, *Frente Nacional de Abogados Democráticos*, quoted in *Proceso*, 19 September 1983). The role of the *Procurador* has been criticised in so far as it only slows the resolution of conflict; eventually, matters must be taken to the Judges for Rental Property (Coulomb, 1985a: 12–13). Similarly, the 1985 decree is deficient in so far as it does not declare contracts void where the landlord demands payments over and above those laid down in the contract. Nor does it cover the situation where small landlords share accommodation with the tenants.

The introduction of the decree in the Federal District had repercussions in several other states. In April 1985, Puebla amended its civil code in a manner which strongly favoured the tenants. The critical element was a prohibition on raising rents by more than 20 per cent annually. At a time when the annual rate of inflation was approximately 60 per cent, it is not surprising that the modification was strongly opposed by landlord interests. The local branch of the Mexican Association of Property Managers (AMPI) argued that the decree would lead to landlords selling out (*El Sol de Puebla*, 16 April 1985); a few days later it threatened that its members would refuse to sign any more rental contracts until a change had been made. This form of rent control, they claimed, was the thin end of the wedge: next there would be a Tenants' Charter.

Attempts were immediately made to calm the landlords. On 22 April it was claimed that the 20 per cent limit only applied to housing and not to businesses. The action on rents had only been taken to protect the 'popular' classes – a duty laid down in the national constitution. By August, however, the Civil Code had again been amended to allow for a single annual rise of up to 70 per cent of the percentage increase in the local minimum salary.[10] Whether the rent controls are being applied in practice is another matter. One current study of the rents registered in the State's Secretariat of Finance claims that the rises are vastly in excess of the rate of inflation.[11]

The whole issue had been highly politicised, the different political

parties arguing their particular case against the backcloth of the forthcoming July elections. The PRI were clearly influenced by the Federal
District's legislation, although some state deputies were recommending
rises lower than the equivalent of 75 per cent of rises in the minimum
salary. On the other hand, two parties of the left, the PST and PSUM, were
proposing a straight 30 per cent rent increase and further legislation to
guard against landlord abuses (*El Sol de Puebla*, 3 April 1985). Elsewhere,
different political processes have again produced different rental codes. No
modifications to the legislation have been made in the State of Jalisco. In
contrast, recent legislation in Michoacán strongly favours the tenant.[12]
   While local government has continued to be sensitive to the issue of
rents, the federal government seems to have omitted rents from its general
anti-inflation strategy. Although most prices and tariffs have been held
constant since 1987, the only major item excluded from the Pact has been
rents. As a result, these have risen rapidly relative to incomes and prices
generally (see Chapter Seven).

## VECINDAD IMPROVEMENT

The quality of living conditions in central *vecindades* has long been a
source of complaint for tenants. An important effect of rent-control legislation has been to discourage landlords from repairing and maintaining
their property. The government authorities have generally done little to
remedy this problem. Isolated decrees and occasional emergency programmes have been approved but little in the way of official funding has
been dedicated to the problem. At the federal level, only INDECO and
INFONAVIT had managed to introduce *vecindad*-improvement programmes before the middle 1970s. More typical was BANOBRAS's policy in
the late 1970s to knock down *vecindades* and to build houses for sale to the
former residents.
   Gradually, however, the protests of tenants' organisations associated
with the different political parties have raised the *vecindad* issue higher up
the political agenda. By the 1980s, rental accommodation was beginning to
reappear in government housing statements; for example, the Federal
Housing Law of 1983 refers to 'the provision of stimuli and help for the
improvement of rental housing' (Article 58). The Law also permits the
establishment of cooperatives to improve and maintain rental housing
(Article 49), the latter forming part of an increasing effort to sell deteriorated property to the tenants. Both FONHAPO and the Federal District
government have begun to encourage the sale of *vecindades* to tenants
(*Unomásuno*, 27 September 1984).
   *Vecindad* improvement figured little higher on the local government

agenda. In the Federal District, for example, although improvement of the housing stock was always claimed to be a major goal, the DGHP did relatively little to improve rental housing conditions. In practice, more *vecindades* were destroyed than improved (Garza and Schteingart, 1978: 128).

Only in Guadalajara did the picture appear more favourable. From the middle 1950s, the State of Jalisco and the Municipality of Guadalajara began to improve conditions in the *vecindades*. Governor Agustín Yáñez, who had used the housing issue prominently in his electoral campaign, established a Directorate for State Pensions (DPE) in 1954. A major responsibility of this organisation was to run a campaign of *vecindad* improvement. It carried out a census of *vecindades* in 1955 and introduced a scheme for physical improvements.[13] Landlords were required to pay for the improvements while the state promised not to raise the cadastral value of the property. Around 30 per cent of the city's 1,600 *vecindades* were improved under this programme. The campaign clearly failed to resolve the problem of deteriorated rental housing, however, for subsequent governors were soon complaining about the housing situation. Some years later, in 1969, Governor Francisco Medina Ascencio denounced the 'intolerable' situation to be found in the *vecindades* (González Navarro, 1974: 225).

In 1975, further efforts were made in Guadalajara. The *Patronato* for the Improvement of Families living in Vecindades was established. It was to inspect properties, draw up plans for improvement, and sign agreements with landlords to carry out the repairs. The owners were to be fined if they did not do the work. Interest-free loans to landlords were made available from a rotating fund established with BANOBRAS money. At the same time, the *Patronato* carried out social-welfare activities in the *vecindades* and provided finance and materials for the tenants to improve their accommodation. There is little doubt that the campaign resulted in a certain improvement to housing conditions. It was also used as a form of propaganda, leaving plaques above the street doors of many *vecindades* throughout the city. At least, the campaign seems to have avoided the Achilles' heel of upgrading programmes, the tendency 'to close deteriorated buildings rather than to improve the housing conditions of the tenants' (Coulomb, 1985a: 12).

It was the effect of the 1985 earthquake that really brought a change to the situation in the *vecindades*, at least in the centre of Mexico City. As we have already seen, however, the expropriation and reconstruction of damaged property was a temporary programme which has not been extended to the majority of residents in Mexico City or to deteriorated rental property elsewhere in the country. The programme demonstrated clearly that the central city lacked sufficient low-income housing, that the existing accommodation ought to be improved, and that the government was able to achieve those goals. However, what it also demonstrated was the continued wish of the Mexican government to convert as many tenants as possible into home-owners.

Indeed, one of the most fascinating facets of the programme was that although the Popular Housing Renovation agency was instructed to conserve the *vecindad* as a way of life, it decided to sell the renovated property to the tenants. The principal reason for this seems to be that the state did not want to be the landlord of the renovated property. Indeed, the morning after the expropriation decree, the head of the Federal District Department declared that the property would be sold to the tenants (Azuela, 1989: 159). While there was some debate as to whether restrictions should be placed upon resale and whether some kind of community ownership would be preferable, the decision not to let the new accommodation was never questioned.

This reluctance by the state to let property is further demonstrated by the follow-up programme of central city rehabilitation called *Casa Propia*. This programme offers loans up to 80 per cent of the cost to groups of inner-city tenants wishing to buy and renovate their *vecindades*. While there are a series of prerequisites, such as the approval of the landlord, income constraints on participants etc., the programme seeks to convert tenants into owners. It is claimed that 25,000 tenants will have acquired homes under this programme during 1988 (de la Madrid, 1988, vol. II: 109).

Alongside these policies in the inner city, state agencies are selling off their rental property. ISSSTE, for example, having tried unsuccessfully to sell off its 21,000 homes in 1984, reinitiated this policy more forcefully in 1987 (*Proceso*, 15 February 1988).[14]

## CONCLUSION

The Mexican state has gradually increased its participation in the housing sector. Between the wars it established several housing institutions but relatively few homes were constructed. After the war, production was stepped up – after 1970, dramatically so. From 1970, the state became a major actor in the provision of housing, a role which was accentuated with the boom in oil revenues. Surprisingly, the recession of the 1980s has seen no decline in the pace of construction – rather the reverse: the production of housing has been a major plank in the government's efforts to maintain levels of employment.

Despite high levels of production, cuts in real incomes and rises in the cost of land and construction have made house purchase difficult for an ever-larger majority. During the 1980s, inflation and the rising costs of credit have effectively priced 'social-interest' housing even further beyond the range of the working class.

Production of housing for rent has played little part in state policy. Apart from a brief period beginning in the middle 1950s, the state has declined to act as a social landlord. Only in the last few years has there been a realisation that more rental housing is needed. Incentives have been offered to the private sector to build for rent, although the results have not been impressive so far. The government seems determined, however, not to build public housing for rent. Its brief experience as a landlord has left a deep mark.

The state's main role in the rental housing field has been as a legislator. The Mexican Constitution delegates responsibility in this field to state governments; they have taken up their responsibility with varying levels of enthusiasm. In some states, rent controls have been applied; in others, little has been done even to maintain the quality of the existing housing stock. The most effective, and controversial, rent controls were applied in the Federal District during and immediately after the Second World War. After the shock of this legislation, however, there was a relative lull in state intervention and it has only been in the past decade that legislative activity has increased. High rates of inflation have again made rent control an active political issue.

In general, therefore, Mexican housing policy has neglected tenants and encouraged the spread of home-ownership. The Federal Government financed increasing numbers of 'social-interest' houses, and the state has also encouraged the development of self-help housing, first covertly, but more recently in an overt and massive way. Today, both strategies are threatened by rising prices and falling incomes.

# 5 Urban development and the housing market in Guadalajara and Puebla

## THE ECONOMIES OF THE TWO CITIES

Guadalajara and Puebla are major regional centres. They are both capitals of their respective states, major industrial centres and the main commercial foci for their surrounding regions. At the same time, there are important differences in the economic traditions and recent development of the two cities. These differences have influenced recent patterns of urban growth and must, therefore, be examined briefly.

Guadalajara was founded in 1532, although the present site was not occupied until ten years later. An Indian uprising in 1540 was fiercely repressed and marked the beginning of a catastrophic decline in the region's indigenous population; compared with other parts of central Mexico, relatively few Indian villages remained (Berthe, 1973: 138–40; Lindley, 1983: 11).[1] In 1560, Guadalajara became the capital of Nueva Galicia, a large area of western Mexico with an *audiencia* which rivalled that of Mexico City. This marked the beginning of what Lindley (1983: 3) describes as the region's 'semi-autonomous' tradition. Nevertheless, Guadalajara grew slowly, its commercial development being limited by poor communications and the trading supremacy of Mexico City (Berthe, 1973). During the late eighteenth century, however, better transport links and a now thriving regional agriculture gave the city new commercial importance. It also led to the establishment of a number of small industries producing leather, textiles, ceramics and soap which began to supply the regional market (Lindley, 1983; Berthe, 1973). Despite this industrial growth, Guadalajara remained essentially a regional centre, participating little in export production (Arias and Roberts, 1985: 153).

At the turn of the century, Guadalajara's industry was still dominated by small, family-run enterprises, a pattern which persisted until the 1960s (Alba, 1986).[2] It was only then that a number of major Mexican and trans-national firms such as Kodak, IBM, Motorola, Celanese, Union Carbide and Burroughs established factories in the city. The new plants were very different from the local enterprises, using advanced technology and employing distinctive management and labour systems. They were attracted to Guadalajara by the large regional market, the favourable urban environment, and the incentives offered by the local authorities (Walton, 1977). Perhaps the most powerful attraction, however, was the city's reputation for peaceful industrial relations, an outcome of highly personalistic worker-employee relations and the traditional tendency for qualified labourers to set up their own workshops in the city's large 'informal' sector (Arias, 1985: 114–17).[3] Today, small-scale activities still employ over half of the manufacturing work-force, and continue to produce 'traditional' consumer products such as shoes and processed foods or drinks (Tamayo, 1982). It is possible that the persistence of this kind of industry explains why Guadalajara has managed to cope better with the economic crisis of the 1980s than many other Mexican cities (Alba, 1986; Escobar, 1988).

Puebla was founded in 1531, in a densely-populated Indian area. The early colonial economy thrived on the basis of wheat and maize production (Liehr, 1976: 15). The Church played an important role in this agricultural development, and by the eighteenth century had accumulated so much wealth that it became the main institutional source of credit (Liehr, 1976).

By the middle of the sixteenth century a textile industry had also been established, producing both wool and silk (Grosso, 1984: 9). Later, when cotton had replaced wool as the major product, Puebla came to be the most important textile centre in colonial Mexico (Bazant, 1977).[4] Together, Puebla's industrial, agricultural and commercial functions made it the second most important city in New Spain, even rivalling, at times, the viceregal capital (Berthe, 1973).[5] By the late eighteenth century, however, Puebla's economic prominence came under threat. The region's agriculture was suffering from labour problems and its industry from competition from other cities (Contreras and Grosso, 1983: 118; Liehr, 1976: 28). Only the cotton industry continued to prosper; from 1835 onwards, local investors established a number of new factories, making Puebla the country's largest textile centre (Gamboa, 1985: 148–50; Aguirre and Carabarín, 1983: 199; Alba, 1986: 101). Control over the industry gradually shifted, however, and by the time of the Revolution a number of Spanish families totally domi-nated the industry (Gamboa, 1985). They, in turn, were displaced by a new group of immigrants, this time from the Lebanon (Alonso Palacios, 1983). During the 1940s, Lebanese immigrant families took over the the industry as well as investing heavily in related sectors. Despite some attempts at diversification, manufacturing in Puebla had become heavily dependent on cotton and by 1960 textiles employed more than two out of every three industrial workers (González, 1980: 72; Mele, 1986a: 8–10). Such a high level of concentration posed a major problem when the increasingly obsolete plants began to face severe competition from plants in other parts of the country. During the 1960s, in fact, a large number of Puebla companies were forced out of business (González, 1980).

Fortunately, a major process of industrial diversification began during the 1960s. The attractions of Puebla's location on the major route between the capital and the country's principal port were reinforced by the completion of the Mexico City-Veracruz motorway in 1962. That advantage, together with government incentives, such as tax concessions and the establishment of new industrial parks, attracted several major new companies to Puebla. Volkswagen opened what was to become the country's largest automobile plant, and Phelps-Dodge, NCR and Ciba-Geigy also established plants in the city.

Although the recession of the 1980s forced most of these plants to cut back on employment and several small companies closed, the crisis appears to have been contained. Indeed, a number of small new plants have been established in recent years (Mele, 1986b).

*Table 5.1* Structure of employment, 1950–80

| | *Percentage of economically active population working in:* | | | |
| | *Agriculture* | *Industry* | *Commerce and services* | *Insufficient information\** |
| --- | --- | --- | --- | --- |
| **GUADALAJARA** | | | | |
| 1950 | 9.6 | 36.5 | 43.5 | 10.4 |
| 1960 | 11.4 | 39.1 | 48.4 | 1.1 |
| 1970 | 5.8 | 39.9 | 48.0 | 6.4 |
| 1980 | 2.3 | 28.9 | 40.3 | 28.4 |
| 1986 | 3.3 | 31.2 | 65.5 | – |
| **PUEBLA** | | | | |
| 1950 | 5.2 | 45.5 | 37.9 | 11.4 |
| 1960 | 15.0 | 37.4 | 47.3 | 0.3 |
| 1970 | 6.8 | 36.7 | 50.6 | 5.9 |
| 1980 | 4.7 | 29.2 | 41.8 | 24.4 |

*Sources:* 1950–80:*Censo General de Población.*[6]
1986:Winnie (1987: 58–9). Estimates based on a survey of 1,937
Guadalajara households.

Even though less than one-third of the labour force is employed in manufacturing and construction (Table 5.1), Guadalajara and Puebla rank among the top four Mexican industrial centres. Both have experienced a major increase in industrial activity since 1950, even if this has failed to maintain the share of secondary workers in either city. As in most other Mexican cities, it is the tertiary sector which has absorbed the bulk of new workers entering the labour force. Allowing for the difficulties with the 1980 Census, it is almost certain that the recession of the 1980s has resulted in a major expansion of this sector.[7]

It is frequently argued that wages in Guadalajara are low by comparison with other cities (Alba, 1986; Escobar, 1986a; Arias, 1985; González de la Rocha, 1986b; Wario, 1984); a phenomenon explained by the city's low level of unionisation, the fragmentation of its work-force, and the person- alised relationships between workers and owners (Tamayo, 1982: 87; Arias and Roberts, 1985).[8] Certainly, a study of industrial wages in 1981 ranked Guadalajara eighth out of twelve major industrial areas – behind cities such as Mexicali, Torreón, Mexico City, Puebla, and Monterrey (Mexico, SPP, 1982).[9] Whereas Guadalajara's industrial workers earned about 10 per cent less than their counterparts in the Federal District or Monterrey, and less than the average for all twelve areas, those in Puebla earned slightly more than the average. Nevertheless, this still constituted an improvement over earlier times. In 1960, per capita monthly income (across all sectors) in Guadalajara was only three-quarters of the average for the sixteen major Mexican cities (Walton, 1978: 39).

A similar pattern is revealed by household income data. In the early 1960s, Guadalajara households received only three-fifths of the average for the sixteen major cities (Walton, 1978: 39), and only four of the sixteen had a larger proportion of low-income households than Guadalajara (Mexico, SHCP, 1964). Interestingly, however, Puebla was one of these four, having a higher proportion of low-income households than any other city but Morelia. For Guadalajara, the situation appears to have improved in the later 1960s and 1970s. Whereas, in 1968, household income in the city was only 48 per cent that of Mexico City, and 76 per cent that of Monterrey, the corresponding figures for 1977 were 84 and 86 per cent (Hernández Laos, 1984: 172). Both household income and consumption were growing faster in Guadalajara than in the two largest cities or any of the eight main regions of Mexico, and as a result the city overtook the prosperous north-western region in this respect during the 1970s. Unfortunately, no disaggregated information on the situation in Puebla is provided.

Although incomes in Guadalajara (and possibly Puebla) seem to have been low at least until the 1970s, incomes were not distributed as unequally as in the other major cities. In the late 1960s, Guadalajara had a much lower

Gini coefficient for income distribution than Monterrey, Mexico City or other large cities (Walton, 1977: 187). A later study also showed that in 1977 Guadalajara had a more egalitarian income distribution than Mexico City, Monterrey or any of the major regions (Hernández Laos, 1984: 172). More recent studies suggest that the distribution of income in Guadalajara changed between 1976 and 1983. While those earning three minimum salaries or more maintained their 18 per cent share, the incomes of poorer households declined dramatically. As Table 5.2 shows, the proportion of households earning less than one minimum salary rose from 8 per cent in 1976 to 40 per cent in 1983.

*Table 5.2* Income distribution in Guadalajara, 1976–83

| *Percentage of households earning:* | | |
| --- | --- | --- |
| | *1976* | *1983** |
| 20 minimum salaries or more | 0.5 | 0.6 |
| 5 to 20 minimum salaries | 9.3 | 10.0 |
| 3 to 5 minimum salaries | 7.9 | 8.0 |
| 2 to 3 minimum salaries | 23.3 | 6.2 |
| 1.5 to 2 minimum salaries | 34.0 | } 35.2 |
| 1 to 1.5 minimum salaries | 16.7 | |
| 1 minimum salary or less | 8.3 | 40.0 |

*Sources:*    1976 – Jalisco, DPUEJ (1979: 108).
1983 – Sudra (1984: 72).

*Note:* * 1983 figures are estimates based on rates of inflation and 1980 Census data.

## POPULATION AND PHYSICAL GROWTH OF THE CITIES IN THE TWENTIETH-CENTURY

Today, Guadalajara and Puebla are Mexico's second and fourth largest cities. Both have grown very rapidly during the past half century (Table 5.3): Guadalajara grew annually by 5.7 per cent between 1940 and 1980; Puebla by 4.6 per cent.[10]

Migration has been a significant component in this expansion although it is difficult to calculate its precise importance because of the deficient way that the Mexican Census classifies migrants. Only those born outside the state of residence are counted as migrants, a clear underestimate since large numbers of city dwellers have been born in other parts of the same state. Officially, therefore, only 21 per cent of Guadalajara's population in 1980 and 17 per cent of that of Puebla were migrants. In contrast, Winnie (1987: 31) claims that one-third of Guadalajara's 1986 population were migrants, and Arroyo (1985: 290) estimates that, at its peak in the 1950s, migration accounted for 67 per cent of population growth. We have no equivalent figures for Puebla.

*Table 5.3* Population of Guadalajara and Puebla, 1900–86 (thousands)

|  | GUADALAJARA | PUEBLA |
|---|---|---|
| 1900 | 101 | 94 |
| 1910 | 120 | 96 |
| 1920 | 143 | 96 |
| 1930 | 180 | 115 |
| 1940 | 241 | 139 |
| 1950 | 401 | 227 |
| 1960 | 812 | 306 |
| 1970 | 1,382 | 513 |
| 1980 | 2,193 | 836 |
| 1986 | 2,888 | 1,120 |

*Sources:* 1900–70 Unikel *et al.* (1976: Table 1-A1).
1980 *Censo General de Población y Vivienda.*

*Notes:* Unikel *et al.*'s calculations refer to the enumeration districts defined as urban within each *municipio* and take account of partial incorporation of surrounding *municipios* into the urban area. The 1980 figure is for the whole *municipio* (including, for Guadalajara, Tlaquepaque and Zapopan).
The 1986 figure is an estimate based on projection of the 1970–80 growth rate.

Rapid population growth has led to major changes in the urban morphology of the two cities. Both have changed from small, neo-colonial, centres into urban agglomerations; both have developed extensive suburbs and industrial zones. In consequence, both have spread beyond their original municipal boundaries (Figures 5.1 and 5.2). The timing of population growth and physical expansion has not, however, been the same for the two cities.

In 1940, Guadalajara was a compact city (Figure 5.1). Peripheral development had been limited and nearby Tlaquepaque and Zapopan were still clearly separate towns. With the population growing at about 5 per cent per year, however, the 1940s saw the first signs of suburban development. By the end of the Second World War, the city occupied 31 square kilometres, compared with its 20 square kilometres in 1930.[11]

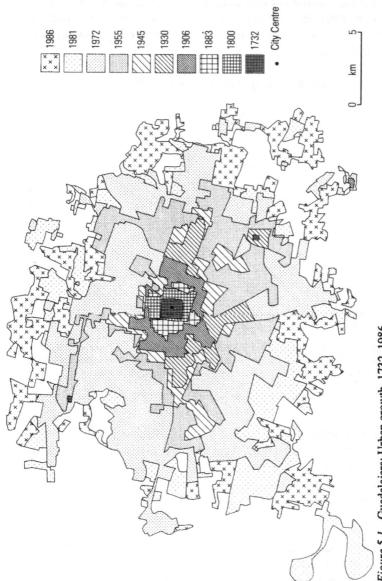

| | |
|---|---|
| 1986 | |
| 1981 | |
| 1972 | |
| 1955 | |
| 1945 | |
| 1930 | |
| 1906 | |
| 1883 | |
| 1800 | |
| 1732 | |
| • | City Centre |

0    km    5

*Figure 5.1*  Guadalajara: Urban growth, 1732–1986

*Sources:* 1732–1906: ITESO, 1984    1930: Preciado and Ibañez, 1930    1945: García Barragán, 1947
1955: Ayuntamiento de Guardalajara, 1955    1972: Maps supplied by Arq. Daniel Vázquez    1981 and 1982: Guía Roji.

City Centre

1984
1977
1974
1965
1950
1930
1900
1804
1698

0    km    5

*Figure 5.2*  Puebla: Urban growth, 1698–1984

*Source*: Mele, 1985.

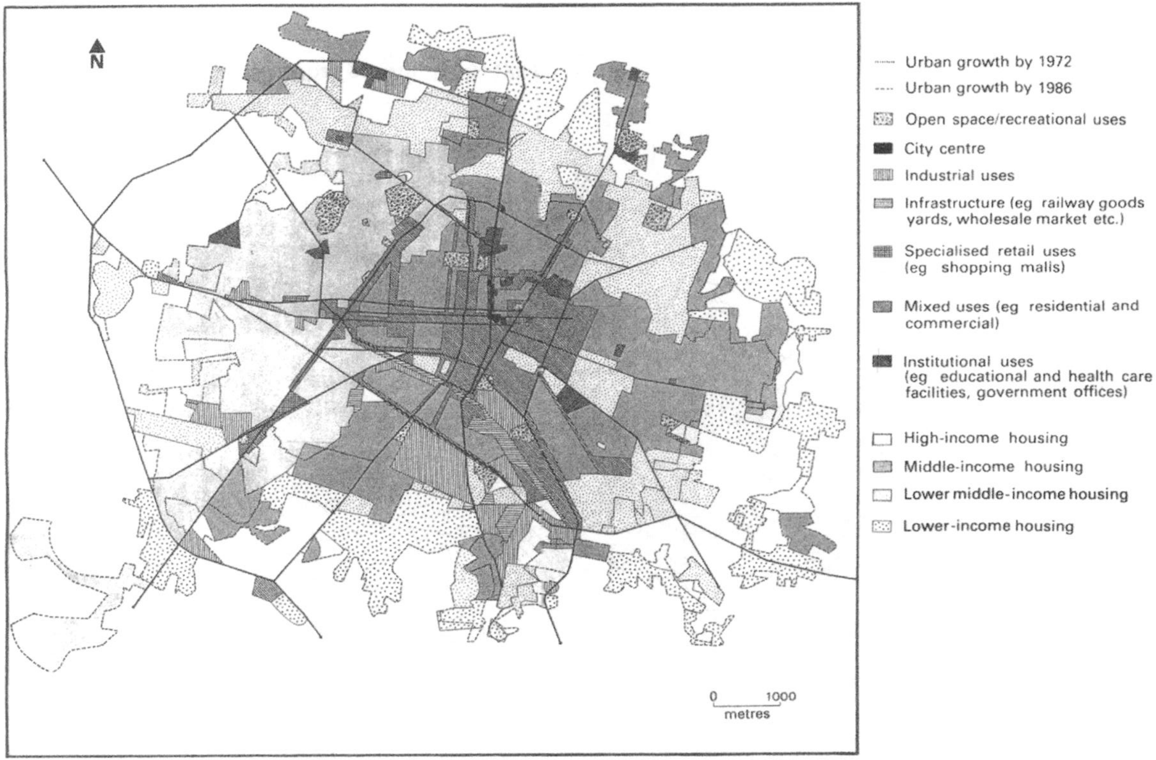
N
Urban growth by 1972
Urban growth by 1986
Open space/recreational uses
City centre
Industrial uses
Infrastructure (eg railway goods yards, wholesale market etc.)
Specialised retail uses (eg shopping malls)
Mixed uses (eg residential and commercial)
Institutional uses (eg educational and health care facilities, government offices)
High-income housing
Middle-income housing
Lower middle-income housing
Lower-income housing
0 1000
metres

*Figure 5.3* Guadalajara: Land use and recent urban growth

*Source:* Adapted from land-use map in DPUEJ/SAHOP, 1981.

The pace of expansion accelerated still further in the late 1940s, when the local authorities attempted to improve the city's infrastructure and to give a major face-lift to the central area. A new railway station and bus terminal were built and several parts of the city were redeveloped and beautified (Walton, 1977: 37). A major programme to widen the streets and to create public squares transformed the city centre, destroying much residential property and encouraging the upper-income groups to move from the central area to newly-established neighbourhoods to the west (Dotson and Dotson, 1953).[12] At the same time, the poor were occupying self-help areas to the east (Figure 5.3). The basis of what Walton (1978) has called the 'divided city' had been laid.

New industrial plants were located in the south and south-west close to the railway and the main road to Mexico City (Figure 5.3). Some working-class neighbourhoods developed in the same area although most poor workers increasingly lived east of the city centre, a pattern of development that has changed little during the years, complicating traffic planning and leading to long journeys to work. The transport system today is unsatisfactory. The private bus service is both slow and overcrowded; the 'metro' consists of trolley-bus routes with underground sections.[13]

The population of Guadalajara grew very rapidly during the 1950s, with an annual growth rate of 7.3 per cent. Expansion of the built-up area also seems to have reached a peak during these years, with the city more than doubling in size in the ten years up to 1955, and again in the following decade (Figure 5.1).[14] In the south-east, Tlaquepaque was gradually surrounded by new suburbs and completely absorbed into the Metropolitan Area. Zapopan suffered the same fate in the 1960s, a decade which saw a major expansion of middle-class areas in the west and north-west of the city (Figures 5.1 and 5.3). Working-class areas grew mainly towards the east and south of the city, although a few emerged in the north and north-west; middle-class areas expanded principally south-westwards around the prestigious Colonia Chapalita.

Despite the declining impact of migration in more recent decades, Guadalajara's population has continued to grow at rates of over 4 per cent per annum. The built-up area has continued to expand quickly, growing from 162 square kilometres in 1972 to 245 square kilometres in 1986. Although high-rise construction has become more common in the last twenty years, the popularity of low-density suburbs has increased. In the south-west of the city, major commercial sub-centres such as the Plaza del Sol have emerged to serve the higher-income, car-owning population. They have contributed to the process of urban sprawl and have given the city an increasingly North American appearance. In addition, low-income settlements have developed in new areas of the city, now forming a virtually continuous periphery around the south, east and north of the city. Urban planning has been unable to contain this process of urban sprawl and has arguably contributed to it.

Puebla has experienced a similar pattern of growth, even if its transformation has occurred more slowly. Certainly, its population was still growing very slowly during the 1930s and it was not until the following decade that the annual growth rate rose to 5 per cent. In 1950, therefore, Puebla was still a physically compact city, occupying an area of only 14 square kilometres (Figure 5.2). Suburbanisation did not really begin until the late 1940s, when planning legislation permitted the development of new areas which were not contiguous with the existing built-up area (Mele, 1986b). The most significant break with the old pattern came with the authorisation of the La Paz estate in 1947. This luxury residential area, located on a hill some distance to the west of the city centre, stimulated the development of other suburbs in the west. A few years later, San Manuel, a less prestigious middle-class area located beyond the urban perimeter to the south-east, encouraged rapid suburban development in that sector of the city. In spite of these developments, physical expansion was still much slower than in Guadalajara.[15]

It was not until the middle 1960s that the urban area began to grow rapidly: between 1965 and 1974 the city's area grew from 23 to 52 square kilometres. Faster suburban growth was linked to changes in residential tastes among upper-income groups and to improvements in transportation, particularly along the main routes to Cholula (to the west), Atlixco (to the south-west) and Tlaxcala (to the north) (Figures 5.2 and 5.4) (Gormsen, 1978). The trend was hastened by public action, for, after 1952, there had been growing agreement among the city's planners that industrial uses should be segregated from residential development (Mele, 1986a: 41). The segregation of land use accelerated as a result of the revitalised growth of industry after 1965. With major new factories locating along the Mexico City–Veracruz motorway, local companies increasingly established facilities in the north of the city. When an 'industrial corridor' containing areas two kilometres each side of the motorway was designated in 1971, and the states of Puebla and Tlaxcala both established industrial estates close to the road, the transition was guaranteed (Mele, 1986a) (Figure 5.4).

Working-class housing estates now began to develop close to the new factories, reinforcing an existing tendency for lower-income groups to live in the north of the city (Gormsen, 1978). The middle-class character of the south and west was reinforced by the development of shopping malls and commercial 'plazas'. Car-ownership was increasing rapidly and major road improvements were being made in these areas, notably the new motorway to Cholula, built in the late 1970s. However, residential segregation never became as marked as in Guadalajara; from the late 1970s, indeed, there was a proliferation of low-income settlement in the south of the city.

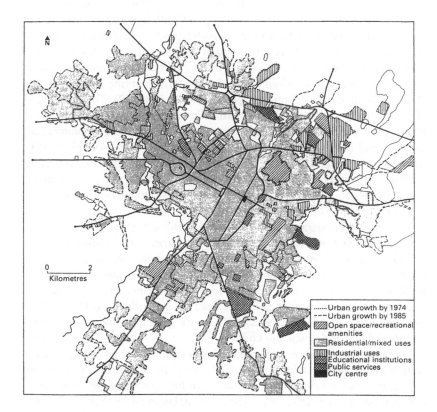

*Figure 5.4* Puebla: Land use and recent urban growth

*Source:* Adapted from Mele, 1985 Maps 2b and 3.

The built-up area probably doubled between 1974 and 1984, and Jones (1989) suggests that it might currently occupy 140 square kilometres. Puebla is currently growing more rapidly than Guadalajara. Figure 5.2 shows that the recent pattern of residential growth has been discontinuous: new housing developments are surrounded by empty land (Mele, 1986b). This pattern, while complicating calculations, has clearly accelerated the spread of the city.[16]

## HOUSING CONDITIONS

In 1950, barely one-fifth of Puebla homes were in owner-occupation; in Guadalajara, three out of ten. Many more households lived in rental accommodation than in cities elsewhere in the country (Table 3.1). One possible reason for this late transition to ownership in Guadalajara and Puebla is the traditional importance of the Church in both cities.[17] As a result, the Church owned a large amount of property in each city, property which was almost always rented out (Bazant, 1977: 10–12). In early nineteenth-century Puebla, the Church owned half the city's houses, accounting for three-fifths of the total value of residential property (Loreto, 1986: 32). The selling-off of Church property from 1856 onwards allowed some tenants to buy their homes, but most of the property was simply purchased by commercial interests and continued to be rented out (Bazant, 1977; Morales, 1985). Renting was the normal tenure of the working class of Guadalajara during the Porfiriato, and property ownership continued to be highly concentrated: in the 1920s, only 14 per cent of the Guadalajara population owned property (Brennan, 1978: 217–38; Vázquez, 1985: 61).

In both Guadalajara and Puebla, the existing tendency for families to rent accommodation was accentuated by the process of industrialisation. Textile companies, important in both cities during the nineteenth century, usually established their plants beyond the urban perimeter. Given the problems of transport the mill-owners found it convenient to build housing, and even to establish a range of related services, for their workers. While this was costly, it had the advantage of capturing not only male workers but also the labour of women and children (Grosso, 1985: 226–27). Although few companies built housing during the twentieth century, the previous pattern had already established the tradition of renting among workers in both cities. As we have seen it was not until the 1950s that the dominance of rental housing began to be eroded in Guadalajara and Puebla, and, even as late as 1980, the majority of Puebla's population continued to rent or share accommodation.

The fact that so many people continued to live in crowded tenements did not seem to worry local elites. While deteriorated rental housing undoubtedly constituted a potential source of problems for the authorities, it was perceived to be less threatening than the proliferation of self-help housing. In Guadalajara the local elite long congratulated itself because the city supposedly faced fewer housing problems than the rest of urban Mexico. Some local academics still accept this conventional wisdom: until recently, Guadalajara 'has proved to be successful in the provision of houses for its workers' (González de la Rocha, 1984: 46). Not surprisingly, the favourable view has been passed on to North American observers visiting the city. For example, Walton (1978: 40) notes 'the unusual absence of land invasions or squatter settlements in Guadalajara' and Logan (1979: 132) comments on the fact that 'in Guadalajara the "shanty-towns" of the poor, which dot the cityscape of other urban areas, are almost completely absent'. Similarly, Handelman (1975: 55) compares the city favourably with the national capital, arguing that Guadalajara 'has more effectively integrated its migrant population into "acceptable" housing'.

Much less has been written about housing conditions in Puebla, which may partially explain the absence of similarly favourable comments about that city. Like Guadalajara, however, Puebla at first sight seems to have escaped the massive self-help housing developments that so 'disfigure' most Mexican cities.

*Table 5.4* Housing conditions in Guadalajara and Puebla, 1950–80

| | | | | Percentage of houses lacking: | | |
|---|---|---|---|---|---|---|
| | *Persons per room *(mean)* | *Percentage of 1(2) room houses* | | *Water*** *Drainage* | | *Electricity* |
| **GUADALAJARA** | | | | | | |
| 1960 | 2.4 | 35.9 | (26.6) | 24.4 | 27.8 | na |
| 1970 | 2.3 | 22.8 | (26.5) | 13.4 | 15.4 | 13.2 |
| 1980 | 1.8 | 13.9 | (24.5) | 11.7 | 8.5 | 5.1 |
| **PUEBLA** | | | | | | |
| 1960 | 2.3 | 42.0 | (22.1) | 26.4 | 29.5 | na |
| 1970 | 1.9 | 30.0 | (24.1) | 24.4 | 23.7 | 15.6 |
| 1980 | 1.7 | 23.0 | (23.0) | 15.5 | 13.9 | 6.1 |

*Source:*   1950–80 *Censo General de Población.*

*Notes:*   *        See note 19.
         **        Census definitions of houses 'with' or 'without' water have been modified as follows: 1950 – houses with a water tank are counted as lacking water; 1970 and 1980 – houses obtaining water from a public standpipe are counted as lacking water. Such houses cannot be disaggregated from the 1960 data.
         na        Not available.

The interesting question is, of course, whether this favourable view of housing conditions in Guadalajara is justified. Is the housing situation in

Guadalajara, and indeed that in Puebla, better or worse than that to be found in other Mexican cities? Calculations by Garza and Schteingart (1978: 28–31) on housing deficits by city, suggest that all was not well in Guadalajara in 1970. Among the fourteen cities with more than 200,000 inhabitants, Guadalajara had by far the worst housing deficit. The city's housing deficit was 62.6 per cent of its total housing stock, compared to an average figure of 46.6 per cent.[18] The same data placed Puebla in a much better light, third best in the listing. Other figures for the same year suggested that 70 per cent of Guadalajara's houses had some physical deficiency, a figure which was high even by comparison with other Mexican cities (Ibañez and Vázquez, 1970).

However, between 1970 and 1980 there seems to have been a general improvement in housing conditions in both Guadalajara and Puebla. Table 5.4 shows that the improvement that was occurring in service provision during the 1960s continued during the 1970s, accelerating markedly in the case of Puebla. In addition, the proportion of single-room houses declined markedly, bring with it a marked fall in room densities.[19]

When housing conditions in the two cities are compared to those in other large Mexican cities in 1980, Guadalajara does appear to be better serviced. A higher proportion of homes are provided with drainage than in any other city listed in Table 5.5, and the city is also near the top of the list for water provision.[20] Puebla performs less well than Guadalajara but is never worse than the overall average and is near the top of the list in terms of drainage provision. In terms of house size, Guadalajara has fewer one-room houses than all but one of the other cities; and also performs well in terms of the number of two-room houses. Puebla has rather more one-room houses than most but is about average in terms of the combined number of one- and two-room houses. It is only in terms of room densities that Guadalajara does not compare favourably: in fact all the cities in the table are similar except those for León, Culiacán and Acapulco, where overcrowding is far worse. The figures in Table 5.5 are not inconsistent with the data presented on tenure structure in Table 3.1. *Ceteris paribus*, cities with higher proportions of tenants might be expected to have more large, consolidated houses and less new self-help construction. As a result, such cities would be concentrated into a smaller physical area which would be easier to service – they should therefore possess superior infrastructure. This argument appears to be partially supported in the two extreme cases of Guadalajara, representing the better consolidated and serviced rental city, and Acapulco, the self-help city.[21]

*Table 5.5* Housing conditions in selected Mexican cities, 1980

| | Percentage of houses with more than 2 people per room* | Percentage of houses with 1(2) rooms | Percentage of dwellings lacking: | | |
|---|---|---|---|---|---|
| | | | Water | Drainage | Electricity |
| Mexico City | 40.7 | 23.9 (25.7) | 10.7 | 10.8 | 1.7 |
| Guadalajara | 36.6 | 13.9 (24.5) | 11.7 | 8.5 | 5.1 |
| Monterrey | 41.0 | 21.4 (24.1) | 12.6 | 20.4 | 4.7 |
| Puebla | 37.2 | 23.0 (23.0) | 15.5 | 13.9 | 6.1 |
| León | 49.2 | 16.6 (17.0) | 15.5 | 23.3 | 10.1 |
| Torreón/G. Palacio | 42.9 | 18.2 (28.1) | 11.3 | 27.9 | 7.2 |
| Ciudad Juárez | 38.1 | 23.9 (28.3) | 9.6 | 27.0 | 8.9 |
| Culiacán | 52.2 | 27.0 (32.9) | 33.2 | 60.7 | 13.6 |
| Mexicali | 37.1 | 18.9 (28.9) | 13.8 | 33.9 | 6.5 |
| Tijuana | 32.9 | 19.6 (28.3) | 29.0 | 33.8 | 8.9 |
| Mérida | 36.0 | 14.9 (34.6) | 27.4 | 38.7 | 4.9 |
| Acapulco | 57.2 | 48.0 (29.3) | 43.8 | 47.3 | 15.9 |
| Chihuahua | 33.5 | 19.2 (21.4) | 12.1 | 23.1 | 8.9 |
| San Luís Potosí | 35.2 | 12.1 (19.9) | 14.8 | 19.5 | 10.5 |
| Aguascalientes | 40.2 | 14.9 (22.9) | 10.0 | 15.4 | 10.1 |
| Toluca | 40.9 | 18.7 (26.8) | 19.9 | 24.6 | 10.3 |
| Morelia | 43.4 | 19.9 (26.8) | 11.5 | 15.7 | 10.9 |
| Hermosillo | 39.7 | 17.5 (25.8) | 13.8 | 33.9 | 10.7 |
| Durango | 40.8 | 19.7 (27.9) | 16.1 | 31.0 | 10.4 |
| Veracruz | 33.6 | 24.2 (26.4) | 10.2 | 13.2 | 3.3 |

*Source:* 1980 *Censo General de Población y Vivienda.*

*Notes:* The cities listed are the same as those in Table 3.1, with the same *municipios* included. Cities are listed in order of their 1980 population (for the entire *municipio*, not only the built-up area if smaller than the municipal area, because housing statistics are given only by *municipio*).

* The Census gives insufficient information to provide a precise figure. Figures given exclude households of 9 or more people living in houses of 5 rooms or more (generally less than 3 per cent, and always less than 6 per cent of houses for which the number of rooms is known).

See notes on definitions in Table 5.4.

## THE NATURE OF THE LAND MARKET

Control over land remains a key element in explaining the tenure structure of most Latin American cities. In Guadalajara and Puebla the form of peripheral land ownership has been a significant influence on tenure. The fact that there has been little public land on the periphery has certainly meant that land invasions have been uncommon. Both private owners and *ejidatarios* have guarded their land against loss. As a result, in Guadalajara, 'wherever land invasions are attempted, the city officials halt them, with the threat of force if necessary' (Logan, 1979: 133); in Puebla, most invasions have been rigorously opposed.

In Guadalajara, invasions occurred with some frequency only in the late 1970s. At that time, a more relaxed government attitude to opposition parties encouraged the Socialist Workers' Party (PST) to promote several invasions, notably in Tetlán on the eastern periphery of the city. However, this strategy was counterproductive because it led to police intervention. Given this reaction by the authorities, the PST received little support from the residents of low-income neighbourhoods (de la Peña, 1988: 19; González de la Rocha, 1986b: 237–9).

In Puebla, recent accounts of social movements in the city since 1960 record at most seven cases of 'successful' invasions (Castillo, 1986; Mele, 1988b). Of these seven, two settlements were demolished, three others relocated, and only in the remaining two were land titles eventually given to the occupants (Mele, 1988b: 19). Clearly, the authorities take strong action against invaders.

In the absence of free land through invasion, the poor have acquired land either through the illegal subdivision of private land or the illegal purchase of *ejido* land. Both mechanisms have been widely used in Guadalajara and in Puebla. However, there have been significant differences in the way the low-income land markets have operated, which may help to explain the differing evolution of housing tenure in the two cities. The chief difference lies in the level of commercialisation: the low-income land market developed much earlier and much more formally in Guadalajara than in Puebla.

### Guadalajara

The characteristics of the land market in Guadalajara were strongly influenced by the early experience of migration during the 1920s and 1930s, a result of major political unrest in the rural areas. The pronounced catholicism of the region's population, and the entrenched power of the Church, led to a counter-Revolutionary movement, the *Cristiada*, breaking out between 1926 and 1929.[22] Peasants fled the rural areas to avoid the violence and the repression that followed the movement's defeat. The presence in Guadalajara of rural migrants from a region in which independence and the ownership of a small plot of land had traditionally been very highly valued meant that, by 1940, there existed a sizeable potential demand for cheap housing plots in the city (de la Peña, 1988: 4; Vázquez, 1985).

Peasants were not the only ones to seek refuge in Guadalajara, however. Landowners and other members of the rural elite also left the rural areas, to escape both the political disturbances and the agrarian reform which was zealously pursued after the *Cristiada* and particularly during the Presidency of Lázaro Cárdenas in the late 1930s.[23] From the 1920s onwards, the region's major landowners had been transferring their investment from the rural areas to the city. In an effort to protect themselves against the effects of the agrarian reform, they started to buy houses, plots and small agricultural properties on the outskirts of the built-up area (Arias, 1985: 85). Their example was followed by small landowners (*rancheros*) and traders moving to the city (de la Peña, 1986: 58). Some newcomers were content to 'live from their rents', but, by the time the agrarian reform put an end to their rural enterprises, others were looking for a new form of investment. Few, however, put their money into industry: the big landowners were conspicuously absent from the ranks of those establishing new industries in the 1930s (Arias, 1985: 85–92; Arias and Roberts, 1985: 158; Walton, 1978: 38; Sánchez, 1979: 252–4).

The agrarian reform of the 1930s led to landowners looking for ways to evade the reform (Varley, 1989a). Large estates close to the city were split into smaller units below the threshold for expropriation; some of these properties came on to the market. Some landowners developed parts of their own property for urban use, to render it physically unsuitable for agrarian reform and to make a larger profit than they would do from selling it as agricultural property. Colonia Atlas, for example, was developed during the 1930s by one major landowner in the south-east of the city.[24]

The process of urban development was seemingly being carefully regulated by the state. Municipal legislation of 1944, superseded by state laws in 1953, 1961, 1969 and 1975, required that the urbaniser gain prior permission from the local council before subdividing the land. Once permission was granted, the urbaniser would install infrastructure and services under the supervision of the authorities. When the facilities were approved, the developer could begin to sell plots. Between 1944 and 1949, thirty-two subdivisions were authorised under the municipal legislation, a major expansion compared with the eighteen new areas which had been developed between 1900 and 1943 (Sánchez, 1979; Vázquez, 1985). The process accelerated over the next twenty years, with 'gold fever' breaking out among developers in the 1960s (Morfín and Sánchez, 1984: 130). By 1975, 175 subdivisions had been authorised in Guadalajara (Vázquez, 1985: 69).

The approval of low-income subdivisions by the Guadalajara authorities helped to create the image of a city which did not suffer from the housing problems besetting other Mexican cities. Walton (1978: 40), for example, comments that 'only 1 per cent of urban residential properties are occupied illegally', and Logan (1984: 41) argues that '*fraccionamiento* developments have been successful in Guadalajara'. What such arguments overlook is the fact that most of the subdivisions were authorised illegally, the result of corruption by the local authorities (Sánchez, 1979; Morfín and Sánchez, 1984; Vázquez, 1985). In practice, most subdividers failed to provide the promised services, and the limited infrastructure that was installed failed to meet the municipal specifications. The authorities registered the subdivision but failed to check whether the subdivider had complied with the legal service requirements. The sale of plots was approved (or simply went ahead) *before* services were installed. In many cases, the municipal authorities were eventually obliged to install the necessary infrastructure with the residents again paying for services which were already included in the price of their plot (Sánchez, 1979).

What is so distinctive in Guadalajara is the apparently routine way in which subdivisions were developed illegally behind a smoke-screen of official approval (Varley, 1989b). That it was effective for so long can be explained in terms of the identity of those involved in the subdivision process. Most were respected businessmen; many were members of various public bodies. Several subdividers were founder partners of the Guadalajara Steel Company; another owned a major local newspaper; others were prominent builders (Sánchez, 1979: 113–15). One subdivider held various positions on the Council for Municipal Collaboration, a body responsible for installing services in many parts of the city, and was later elected to municipal office with responsibility for public works (Vázquez, 1985: 71–4; Sánchez, 1979: 113–19). Other subdividers were founder members of the local Chambers of Commerce and Construction (Sánchez, 1979).

The subdivision process in Guadalajara is well illustrated in the case of *colonia* Agustín Yáñez (a settlement where we carried out interviews). The main subdivider was a man responsible for over 70 per cent of the low-income subdivisions in the east of the city (Sánchez, 1979). He owned several construction companies and a number of radio stations. His partner in Agustín Yáñez was a property dealer and builder involved in public-works projects. They applied for permission to subdivide in 1950, at the same time as they were arranging to purchase the land from a man who had bought it six years earlier.[25] In spite of doubts expressed by the planning department, the Municipal President 'provisionally' authorised the sub-division and occupation of plots: sales could go ahead even though the authorities were aware that the area lacked services. By 1962, four-fifths of the plots had been occupied, but there was still no street paving, pavements, water, drainage or street lighting, services which were not finally installed (at the residents' expense) until 1971.

Clearly, the reality of the subdivisions such as Agustín Yáñez was very different from the rosy image sometimes presented of Guadalajara's low-income housing up to 1970. However, the process was relatively 'formal' and 'commercialised', with the subdividers trying to adapt a system that worked for middle-class housing to low-income housing. They were both unscrupulous and professional in their approach and were responsible for large numbers of subdivisions.[26] Increasingly, over the years, they were able to turn to the commercial banks for loans to finance their projects (Sánchez, 1979: 247). They were sufficiently close to the local authorities that they did not have to fear the consequences of non-compliance with the legislation. On the other hand, they did go to the trouble of seeking permission for their developments, and this sometimes involved them in making payments which they might otherwise have avoided. This may indicate a need to comply with the spirit, if not the letter, of the *tapatíos'* preoccupation with 'city planning'; it was probably also good for business in so far as it helped to deceive the purchasers.

By the late 1960s, however, it was becoming increasingly obvious that the model was working less effectively, in part at least because the sub-dividers were attempting to squeeze too much out of their developments (Morfín and Sánchez, 1984: 137). This is most clearly seen in Santa Cecilia, an area of north-east Guadalajara developed by the same subdivider as in Agustín Yáñez (Logan, 1984; Sánchez, 1979).[27] This neighbourhood was particularly poorly serviced and the streets were only 5 metres wide. Plot sizes were reduced to a mere 75 square metres, well below the legal minimum of 90 square metres, and purchasers who got behind with their payments were ruthlessly treated.

Santa Cecilia was virtually the last of the low-income subdivisions. Their demise was a product of several factors but clearly the rising cost of land and servicing was a critical component (Morfín and Sánchez, 1984: 137). Better services were required because of national directives from President Echeverría and because protests had broken out in low-income areas such as Santa Cecilia against the abuses by the subdividers (de la Peña, 1988). In 1973, the newly-installed mayor declared the 'popular subdivisions' to be fraudulent and insisted that subdividers provide the infrastructure required by the law. The result of this insistence was both to slow the development of the subdivisions and to raise the price of the plots. After 1976, when new legislation controlling subdivisions was approved, the 'popular subdivision' ceased to be a profitable activity for property developers, who moved into middle-class housing construction, particularly in the west of the city (Morfín and Sánchez, 1984: 132).[28]

From the early 1970s, therefore, the process of land acquisition for low-income housing in Guadalajara began to change. The change was facilitated by the fact that the fringe of the expanding city was now contiguous to the extensive area of surrounding *ejido* land (Figure 5.5). Henceforth, a series of *ejido* subdivisions began to replace the 'popular subdivisions'. Increasingly, land was sold by *ejidatarios*, and particularly by their local leaders, to low-income settlers. According to the Agrarian Reform Laws, this process was illegal, but it was facilitated by official and semi-official connivance. Most of the *ejido* subdivisions depended on illicit activity on the part of officials of the agrarian reform ministry and related agencies (Sánchez, 1979: 275).[29] However, compared with the situation in Mexico City, influential outsiders seem to have been more involved in *ejido* land sales in Guadalajara (Sánchez, 1979: 250). For example, an important trade union leader and former mayor of Guadalajara was responsible for establishing several settlements on *ejido* land in the vicinity of Santa Cecilia (Universidad de Guadalajara, Instituto de Asentamientos Humanos, 1985).[30] Outsiders were also involved in the south of the city. In Lomas de Polanco, the developer illegally took over land from two *ejidos*, in the face of *ejidatario* opposition, to carry out his scheme (Morfín and Sánchez, 1984). In Buenos Aires, plots were sold directly by *ejidatarios*, but here, as in neighbouring settlements, the stimulus to sale came from leaders of the National Peasant Confederation (CNC). Individual *ejidatarios* were promised 50 per cent of the sale proceeds and persuaded to sign the documents 'ceding' land to the purchasers.[31] Like the subdividers of private lands, the group behind the development of Buenos Aires benefited from close links with the municipal authorities, the police being used to evict dissident settlers.

The involvement of non-*ejidatarios* in land sales in Guadalajara is also apparent in the development of *ejido* lands for middle-class housing. In the 1950s, a number of private individuals and companies manipulated legal provisions for the 'exchange' of *ejido* and private lands in order to build in north-eastern Guadalajara (Vázquez, 1989). Luxury housing was built on *ejido* lands near the University's School of Architecture on the edge of a spectacular canyon to the north of the city. More recently, *ejidatarios* in the west of the city tried to sell their land to a property developer for middle-class housing and, when this failed, the area was used for 'social-interest' housing.[32] Wario (1984: 161) estimates that some 30 per cent of *ejido* land

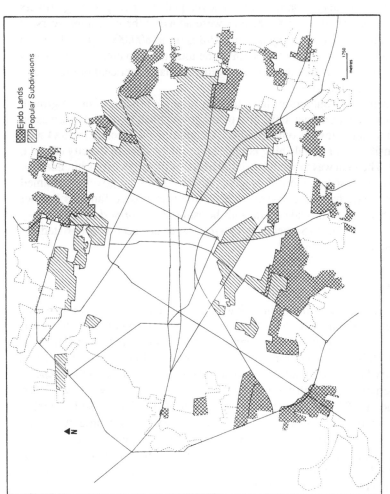

*Figure 5.5* Guadalajara: 'Popular' subdivisions and urbanised *ejido* lands

*Source:* Adapted from maps supplied by Arq. Daniel Vázquez, and Jalisco, DPUEJ, 1979.

developed in the city has been used for higher-income settlements.

The importance of *ejido* areas in the development of the city is demonstrated by Sánchez' (1979: 276) estimate that, between 1970 and 1975, two-fifths of newly-urbanised land occupied such areas. By 1982, some 3,500 hectares of *ejido* land had been developed and local planning officers estimate that another 1,300 hectares were added in the next three years. If these figures are correct, by the mid-1980s up to 800,000 people lived on such land, approximately one-sixth of the total built-up area.[33] In the south of the city, local planning documents suggest that 56 per cent of urban land once belonged to *ejidos*.

Recently, the authorities have tried to slow the process of urban growth by stricter control. They have also begun to establish territorial reserves on which to house low-income families. A committee was set up in 1984 with responsibility for establishing land reserves on the edge of the city (Zavala, 1984). The plan was to establish 2,100 hectares of subdivisions intended for popular and 'social-interest' housing, and by 1986 the governor claimed that the state had enough land to accommodate 60,000 families but not enough funds to provide them with housing (*El Informador*, 2 February 1986). By 1989, 3,151 hectares of *ejido* land had been expropriated for legalisation.[34]

## Puebla

In Puebla, the difficulty of invading land also stimulated the emergence of the illegal subdivision. The form that this development took, however, shows several significant differences from the process in Guadalajara.

Puebla's traditional importance as an industrial centre has meant that property has been less attractive to the city's investors than to their *tapatío* counterparts. Although many owners of textile firms also owned *haciendas*, Gamboa (1985: 212) notes that only eleven of the sixty or so key textile owners in the early twentieth century had invested in *urban* property. When the Lebanese began to challenge the city's established economic elite, they did so by investing in textiles and other industries. It was not until the 1960s that the Lebanese industrialists became involved in construction and property development (Mele, 1986a: 10).

Compared with Guadalajara, Puebla was more actively involved in the Revolution. Although the city was fiercely Catholic and right-wing, the states of Puebla and Tlaxcala actively supported Zapata's peasant rebellion.[35] Consequently, both States were among the first to experience extensive land redistribution, 'a reward and inducement for the continued support of the federal government by peasants' (Sanderson, 1984: 81).[36] Thus, whereas most *ejidos* around Guadalajara were not founded until the

1930s, the earliest Puebla *ejidos* date from 1918. Further grants were made during the 1920s and early 1930s, so that most of the land grants made under Cárdenas were extensions of existing *ejidos* (Méndez Sainz, 1987).[37]

Unlike their Guadalajara counterparts, therefore, most of the *ejidos* around Puebla were formed before the city's population began to expand rapidly. Indeed, the earlier pattern of agrarian reform in the region is likely to have delayed urban growth by reducing migratory pressures. Consequently, when landowners in Puebla were coming under renewed pressure due to the Cárdenas reforms there was less demand for new housing land. In contrast to the situation in Guadalajara, the pace of urban growth did not require the rapid conversion of rural into urban land.

As a result, landowners seeking to escape the reform around Puebla divided their land into smaller agricultural properties. The large landowners did not divest themselves entirely of their former property. Indeed, apart from the Lebanese, the city's major landowners are still members of the industrial and political elite who bought their land 'long before the rapid growth of the urban area' (Jones, 1989).

When population growth did accelerate in the 1940s, the process of land alienation in Puebla was rather different from that in Guadalajara. This process is well illustrated by the case of Veinte de Noviembre.[38] The settlement was founded on land belonging to two small *ranchos*, both of which were subdivisions of much larger estates. Permission to develop part of Veinte de Noviembre was granted in 1944; the other part was sold without authorisation. The land was subdivided by four intermediaries acting on behalf of the owners. The daughter of one of these intermediaries recalls that her father approached the owner of the larger ranch to see if he could buy a plot of land to build a house. The owner said that he was no longer interested in working the land, and that they could subdivide it for him if they were interested in doing so. In return, the four intermediaries were each given control over various plots; the ranch owner merely signed the 'contracts of sale'. As many residents recall, the area remained semi-agricultural for years: isolated houses were surrounded by fields and cattle and residents had to wait many years for services. Unlike their Guadalajara counterparts in Agustín Yáñez, the ranch owners (and intermediaries) continued to live in the area; their relatives were still living there in 1986.

The state of Puebla passed a subdivision law in 1940, a decree that remained in force until superseded by a new law in 1974. However, relatively few subdivisions were ever registered, apparently the result of indifference on the part of the authorities. Up to 1960, only 17 per cent of subdivisions had been authorised (Mele, 1988b) and even in the early 1980s only 96 out of 301 settlements in the city had been developed legally (Méndez Sainz, 1987: 32).[39] Extensive archival and cartographic work by

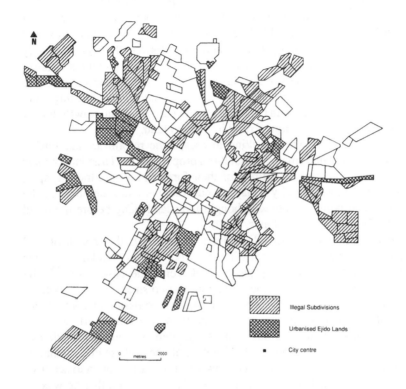

*Figure 5.6* Puebla: Illegal subdivisions and urbanised *ejido* lands
*Source*: Mele, 1988b.

Mele (1988a) shows that by 1986 there were 70 officially-approved private subdivisions, compared with 154 illegal ones. Although the state and municipal authorities have been more zealous in registering subdivisions in recent years, it is interesting to note that unauthorised subdivisions of private lands were still occurring in the city in the 1980s.

The character of the developers is also different from that in Guadalajara, particularly in terms of the scale of their activity. In Puebla most subdivisions have been developed by small-scale investors. While subdividers have included members of the city's economic and political elite, notably a former state governor who developed three authorised subdivisions and a former municipal president and industrialist who founded the upper-income suburb of La Paz (Mele, 1988b: 17–18), there is a far less systematic involvement in property development than in Guadalajara.[40]

As in Guadalajara, however, the occupation of *ejido* land has become increasingly important over the years (Figure 5.6).[41] Mele (1984: 6) estimates that, since 1955, 49 per cent of new settlement has been organised in private subdivisions, 14 per cent in public housing projects, and 37 per cent in illegal developments on *ejido* land. Since 1970, 46 per cent of urban growth has occupied *ejido* land, 17 per cent has taken the form of public housing projects, and only 38 per cent has been accommodated through the subdivision of private lands. In 1983, the mayor of Puebla recognised this trend, estimating that 40,000 residential plots had been developed on *ejido* land (Mele, 1987: 24), and in 1986, the land regularisation agency (CORETT) estimated that the number was nearer to 50,000 plots. In contrast with the situation in Guadalajara, the sale of *ejido* land has been mainly the responsibility of *ejidatarios* and their leaders, aided by agents or officials of the Agrarian Reform Ministry.[42]

As we noted above, it is not only the poor who occupy *ejido* land. In Puebla, even public buildings (such as the municipal fire station) have been constructed on such land. In general, the Mexican state has responded to the illegal occupation of *ejido* land less through controlling the process than by subsequently regularising such developments (Varley, 1985a; 1985b). Legalisation is a complicated and highly politicised problem; and the bureaucratic procedures involved also hinder the process. Progress in Puebla has been particularly slow. By 1986, only 8,000 or so plots had been mapped prior to attempts at regularisation; and by 1989, only 110 hectares, belonging to four *ejidos*, had been compulsorily purchased prior to regularisation. In more recent years, the state has also tried to slow the pace of illegal development. In 1983, following a national initiative, a State of Puebla Land Reserves and Legalisation agency was established, one of whose functions was to create reserves for 'social-interest' housing. The authorities have also announced their intention of turning sixteen *ejidos* into enterprises to legally urbanise land for low-income families (Castillo, 1986), although it is doubtful whether this aim will be achieved.

## TRENDS IN HOUSING TENURE AND THE NATURE OF THE LAND MARKET

Table 5.6 shows that the percentage of owner-occupied housing declined marginally between 1950 and 1960 in both Guadalajara and Puebla. This was the period of peak population growth in Guadalajara, although growth in Puebla's population had actually slowed down. In both cities, however, the creation of new homes failed to keep up with the pace of migration, and housing densities increased markedly in both cities. In Guadalajara, even though the number of homes increased by over 5 per cent annually, housing densities rose from 4.9 persons per house to 5.9. In Puebla, despite much slower population growth, the number of homes was increasing at only 2 per cent per annum, and densities rose from 4.8 persons per house in 1950 to 5.6 in 1960. At this time, therefore, it appears that neither the rental sector nor informal owner-occupation were able to cope with demand: neither were creating housing units in sufficient numbers to keep up with population growth.

*Table 5.6* Tenure, population and housing density, 1950–80

|  | Annual population growth (%) | Percent-age owned | Annual growth in housing stock (%) | | | Persons per house |
|---|---|---|---|---|---|---|
|  |  |  | Total | Owned | Rented/non-owned |  |
| **GUADALAJARA** | | | | | | |
| 1950 | 5.2 | 30.9 | – | – | – | 4.9 |
| 1960 | 7.3 | 29.9 | 5.4 | 5.1 | 5.6 | 5.9 |
| 1970 | 5.5 | 45.7 | 4.2 | 8.7 | 1.6 | 6.7 |
| 1980 | 4.7 | 52.1 | 6.1 | 7.5 | 4.8 | 5.5 |
| **PUEBLA** | | | | | | |
| 1950 | 5.0 | 20.8 | – | – | – | 4.8 |
| 1960 | 3.0 | 20.6 | 2.0 | 1.9 | 2.0 | 5.6 |
| 1970 | 5.3 | 38.7 | 4.8 | 11.7 | 2.1 | 5.6 |
| 1980 | 5.0 | 47.5 | 5.2 | 7.4 | 3.6 | 5.3 |

*Sources:*  Population growth:  Unikel *et al.* (1976: Table 1-A1); *1980 Censo General de Población*; and own calculations.
Other statistics:  1950–80 *Censo General de Población*.

*Notes:*  Annual growth rates refer to the average percentage increase over the previous decade.
Different criteria are used for housing tenure in the Census for different years. In 1950 and 1970, houses were described as owned or non-owned; in 1960, as owned or rented; and in 1980 as owned or rented, but with an unidentified third category. The possibility that other types of tenure have been included under different categories in the figures for 1950–70 may account for some of the variation observed. The variation in criteria also make it impossible to quote a figure for the proportion *rented* over the whole period.
For *municipios* included, see note 6; but note that the population growth figures refer to urbanised enumeration districts within the *municipio*, not the whole *municipio* (see Table 5.3).

This trend continued during the next decade. In Guadalajara, growth in the housing stock fell to just over 4 per cent per annum, lagging behind population growth and leading to a continued rise in housing densities. In Puebla, housing construction accelerated but was still unable to keep up with demand. However, the number of owner-occupied homes grew dramatically in both cities, increasing by 9 per cent per annum in Guadalajara and 12 per cent in Puebla. The effect on the tenure structure was obvious: the share of the housing stock in owner-occupation rose dramatically in both cities, nearly doubling in Puebla. In this sense, the experience of the two cities was very similar.

During the 1970s, the rapid transition to ownership continued in both cities. However, the number of homes in non-ownership had begun to grow more rapidly so that housing densities at last began to decline in Guadalajara and, less rapidly, in Puebla.

To what extent can we account for these changes in terms of our earlier discussion of the nature of the land market in each city? One of the most striking features of Table 5.6 is the rise in housing densities in both cities, during the 1950s, and, in Guadalajara, during the 1960s. The creation of new homes simply failed to keep up with demand. Given the normal recourse to self-help construction in most Latin American cities, this failure suggests that during the 1950s, land was not being made available in sufficient quantities to low-income people. Admittedly, the housing stock in Guadalajara was increasing annually by over 5 per cent, but this was clearly insufficient to accommodate a population which was increasing annually by more than 7 per cent. The so-called 'legal' subdivisions were providing opportunities for home-ownership, but not in sufficient numbers. Those who could not gain access to a plot were forced to crowd into the existing housing stock. In Puebla, where there was no real expansion in self-help housing at this time, a mechanism for making cheap plots of land available to lower-income people in significant numbers had simply not emerged. As a result, new migrants were obliged to crowd into the existing accommodation. Fortunately, the pace of population growth was much slower in Puebla than in Guadalajara.

It was only in the 1960s that, with increasing access to *ejido* land, the pace of self-help ownership could accelerate in both cities. Home-ownership increased dramatically: in Puebla the number of owner-occupied homes was expanding annually by almost 12 per cent. Nevertheless, the pace of this expansion was still insufficient to reduce housing densities. With the 'legal' subdivisions becoming more expensive, some of the poor in Guadalajara could ill-afford to become home-owners even though the owner-occupied housing stock was growing at nearly 9 per cent per annum. Our argument that the market for low-income land in Guadalajara was more commercialised than that of Puebla is highly relevant at this point. It will also be recalled that the business of 'legal' subdivision for low-income families started to come under pressure towards the end of the decade.

It was in the 1970s, however, that the data suggest most clearly that land was again becoming scarce. For, while housing densities were falling slightly, the rate of growth of non-ownership began to rise. While home-ownership through self-help settlement was clearly still possible, the conditions facing potential self-help owners were dissuading many from taking this option. Unfortunately, we cannot confirm whether cost was the major factor, because there are no studies of the economics of low-income housing at this time and certainly no adequate data on the cost of land. We cannot be certain, therefore, whether it was the cost of land that was the principal barrier to home-ownership in self-help settlements.

What we can say is that when population growth was at its peak, during the 1950s in Guadalajara and the 1960s in Puebla, the land market accommodated this expansion by making plots available for owner-occupation. On the whole, these plots were sufficiently large to satisfy the demand of most aspirant home-owners. The average size of plots in both cities suggests that neither land market was especially tight during these years. It was only in Guadalajara in the early 1970s that the average size of plot fell dramatically; in the problematic settlement of Santa Cecilia plot sizes averaged only 90 square metres compared to at least 150 square metres in earlier subdivisions. This difficult period was overcome, however, by the increasing occupation of *ejido* lands and plot sizes again began to grow.

## THE CURRENT PRICE OF LAND

Of course, what we most need to support these kinds of argument are data on the cost of land relative to income in different years. Unfortunately, such data are scarce and we are only able to present information for the early 1980s based on the data collected in our survey. Table 5.7 contains the mean and median prices per square metre in the young settlements during the three years in which most plots were sold. These prices are compared with the minimum official daily wage in force at the time. The data on median price show that one minimum salary bought approximately two square metres of land in the Guadalajara settlement in 1980 and little more than one square metre two years later. In the Puebla settlement, a daily minimum wage bought approximately one square metre of land in 1983 and somewhat less than that in 1984. The fact that the prices are rising relative to the minimum salary is partly the effect of the falling real value of the minimum wage and partly the effect of rising prices in the two settlements due to growing levels of settlement consolidation.

We believe these two settlements to be reasonably typical of peripheral land prices in the two cities at the time and the figures are broadly confirmed by land price data provided in confidence by public officials.[43]

*Table 5.7* Cost of land in survey settlements compared to minimum salaries (current prices for year of purchase)

| Year | Mean plot size | Plot price per $m^2$ | | Minimum salary | Mean price per $m^2$/min. salary | Sample size |
|------|------|------|------|------|------|------|
| | | Mean | Median | | | |
| GUADALAJARA (Buenos Aires) | | | | | | |
| 1980 | 130 | 109 | 81 | 145 | 0.75 | (20) |
| 1981 | 140 | 139 | 93 | 190 | 0.73 | (21) |
| 1982 | 122 | 239 | 179 | 255 | 0.94 | (16) |
| PUEBLA (El Salvador) | | | | | | |
| 1982 | 284 | 362 | 279 | 225 | 1.61 | (16) |
| 1983 | 250 | 429 | 358 | 393* | 1.09 | (18) |
| 1984 | 218 | 965 | 719 | 605* | 1.60 | (14) |

*Notes:* * The minimum salary changed on 1 January and on 1 July. The figures given here are an average of the two figures.

Another way of evaluating the cost of land to the poor is by comparing the cost of purchase in a peripheral settlement with monthly rents. Table 5.8 estimates that in Guadalajara it would take between 41 and 45 months' rent in a *vecindad* to buy a typically sized plot of land. Compared to the rent of a flat in an older self-help settlement, it would cost only 21 months' rent. In Puebla, the costs of purchase seem somewhat higher but there is greater variation in the level of *vecindad* rents. Taking the lower rent level, it would cost 66 months' rent to buy a plot. On the other hand, the *vecindad* rents recorded in the older self-help settlement would buy a plot in 38 months. In both cities calculating on the basis of the median plot price would reduce the respective figures.

On the basis of this limited information we can say that although prices are much higher than, for example, in Chihuahua (Hoenderdos, 1985), they do not seem to be out of reach of the poor. It is very clear, however, that marshalling the funds to put down a deposit on a plot of land is not easy, especially under current economic conditions. We consequently have little doubt that many families who are prepared to face the burdens of self-help construction will be unable to take advantage of this tenure option. We return to this issue in the next chapter.

*Table 5.8* Land price in peripheral settlements relative to rent levels (1985 prices)

|  | GUADALAJARA | PUEBLA |
|---|---|---|
| *Vecindad* rent (central city) | 3,500 | 2,700 |
| *Vecindad* rent (older settlement) | 3,200 | 4,700 |
| Flat (older settlement) | 6,800 | 6,300 |
| *Approximate average price of land per square metre*: | | |
| Mean | 1,200 | 1,500 |
| Median | 900 | 1,150 |
| *Cost of 120 square metre plot*: | | |
| Mean | 144,000 | 180,000 |
| Median | 108,000 | 138,000 |

*Notes:*  For definitions of *vecindades* and flats, see Table 6.4 and note 7 in Chapter Six. The rents are those quoted by tenants interviewed in the questionnaire survey. The cost of land is that paid between 1980 and 1982 in Buenos Aires and 1982 and 1984 in El Salvador, expressed in 1985 prices. It would not have been possible to buy a plot of 120 square metres in El Salvador as plots in that settlement were typically much larger. However, the same figure has been used for Puebla and Guadalajara for ease of comparison.

One other issue should also be raised. Much of the housing literature suggests that land prices in Latin American cities are rising rapidly. Certainly, there is little doubt that land purchase is frequently seen to be a lucrative investment and that much commercial capital is put into land. This has consistently been the case in Guadalajara and Puebla, where it has resulted in the emergence of a great deal of vacant land within the built-up area (Mele, 1988b; Wario, 1984). In both cities, wealthy speculators buy large tracts of land on the periphery while smaller investors buy two or three plots in newly-developing settlements. This process has given rise to claims, in Puebla, that one-fifth of land within the urban perimeter was vacant in 1980 (Puebla, 1980) and, in Guadalajara, that 15 per cent of all plots were empty in 1984 (*El Informador*, 7 January 1984).[44]

In the absence of any study of this phenomenon in either city, we collected information on land prices as recorded in advertisements in the cities' main newspapers during 1975, 1980 and 1985.[45] In Guadalajara, this yielded a total sample of 446 plots but in Puebla, where estate agents are a more important source of land sales, this procedure was much less effective. In neither city were the data suitable for measuring price changes in poorer settlements. Nevertheless, the price per square metre for each advertised plot in Guadalajara was calculated and the prices converted to 1985 values. An unweighted average was then taken for all prices in each year. Contrary to our expectations, there was little sign of rampant price inflation. The average price per square metre certainly rose between 1975 and 1980 but only by 14.4 per cent. In contrast, between 1980 and 1985 the average price fell by 3.1 per cent.

Of course, the locations of plots being offered for sale in 1980 were different from those of plots advertised five years later; although, in view of the strong outward spread of the city, there was remarkably little change. However, in order to guard against the danger of comparing the prices of totally different areas of land, we considered the changes in price on a settlement by settlement basis. We have data on changes in land prices between 1975 and 1980 for fifteen settlements. Of these fifteen cases, only one showed a real price decline. In contrast, the data on prices in twenty settlements in 1980 and 1985 reveal fourteen cases where prices rose and six where they fell. The largest single rise in price between 1975 and 1980 was 86 per cent, between 1980 and 1985, 53 per cent. While these are substantial rises, we cannot be entirely sure that they are for comparable kinds of property. In any event, the rises hardly support the kinds of dramatic statement that usually accompany discussions of land speculation.

The usefulness of these data is limited, of course, because the information is confined to the one city and is also generally restricted to the western quadrant of Guadalajara, an area where there are few poor families. While evidence from Bogotá suggests that price trends in low-income areas rise in broadly similar ways to those in higher income areas (Mohan and Villamizar, 1982), we cannot draw too many conclusions about price trends in the city.

## CONCLUSION

Guadalajara and Puebla have grown rapidly during the past fifty years, their populations more or less doubling every fifteen years. Economic growth has been sustained by a combination of industrial, commercial and administrative expansion. Both cities combine employment in major manufacturing plants with large numbers of small-scale industrial, commercial and service activities.

For many years, most of the population was housed in rental accommodation. The cities long remained physically compact and it was only with improvements in transportation and changing residential tastes that the situation began to change. Eventually, suburban development allowed many middle-class families to become home-owners and many poor families to gain access to a home. For the poor, however, the transformation from tenant to home-owner came more slowly than in most other Mexican cities. In 1960, most of the poor were still tenants.

Eventually, self-help construction became a common strategy among the poor of both cities even if land in Guadalajara and Puebla was rarely obtained through invasion. Of course, there were differences between the two cities. Guadalajara began to grow earlier than Puebla, which encouraged the emergence of a more 'commercialised' model of subdivision of land development for the poor. Many investors, in the process of shifting funds from the rural sector, found an attractive outlet in the property sector, at the same time as lands were becoming available for urban development as a result of landowners near the city seeking to evade the agrarian reform. In Puebla, industry attracted more funds and property development was less commercialised than in Guadalajara. The fact that agrarian reform preceded rapid urban growth meant that landowners in the area around the city had less opportunity to convert their land into urban real estate.

In neither city did the housing stock expand as rapidly as population until the 1970s; housing densities rose in both cities during the 1950s and in Guadalajara continued to do so during the 1960s. The high costs of land acquisition in Guadalajara and the less dynamic process of subdivision in Puebla slowed the transition to ownership. Renting remained important longer than in other Mexican cities.

Eventually, however, the physical expansion of the two cities led to a major change in the dominant mode of land subdivision. Increasingly, the built-up area encroached onto *ejido* land. Given that *ejidatarios* found land sales more profitable than farming, *ejido* land entered the market in increasing quantities. The shift of residential development from private to *ejido* land increased access to land for the poor. What in Guadalajara had been a relatively expensive option and, in Puebla, one that was not common, came within the means of larger numbers of poor families. Extensive areas of *ejido* land remain, which, despite speculation, has kept price increases within bounds. A plot of land is certainly not cheap in either city, but still seems to be affordable. Tenants can acquire a plot for the cost of renting a *vecindad* room for three or four years.

Saving money to buy land is not an easy task in recession-hit Mexico. There have been severe cuts in income, especially for those receiving the minimum salary. The affordability of land is also threatened by the activities of the authorities to control the pace of urban sprawl. In an effort to slow urban expansion and ease the task of service delivery, they are making some attempt to restrict illegal land sales. Their success in achieving this goal will help determine whether the now well-established shift from tenancy to ownership can continue in the two cities.

# 6 Residential tenure: choice or constraint?

For most households the choice of residence takes place in a highly constrained environment. Few live in the kind of housing that they would wish to occupy. Clearly, the constraints on the poor, on racial minorities, and on single-headed households are greater, but all households face some constraint in their choice of a home. These constraints include the cost of housing relative to income, the location of housing relative to work, the quality of the stock of housing available in any city, and so on. The intention of this chapter is to examine the process of tenure selection in order to identify the main constraints on residential choice in Guadalajara and Puebla. We begin with a review of the findings of the general housing literature before turning to the results of local surveys.

## LIFE-CYCLE EXPLANATIONS

Most studies of residential tenure in the cities of developed countries link tenure choice to the nature of the household. Age, size of household, and structure of the family group are regarded as critical elements determining tenure choice. Such an approach is justified in so far as household characteristics clearly affect tenure choice. In most societies, young single households choose different kinds of accommodation from those selected by retired couples; families with young children make different tenure choices from those of childless couples. In terms of location, some households want suburban housing, some central locations, some access to work, others access to beaches, parks or rivers. In terms of space, some households need a lot of room, while others have different priorities.

In most developed countries, as well as in an increasing number of less-developed nations, these different sets of preferences are clearly related to life-cycle changes: the young live with parents; young single adults rent rooms or flats; families with children tend to move into larger flats or houses; elderly people tend to move into smaller homes. Different stages in the life-cycle generate different sets of residential needs.

Over and above the issue of residential needs, however, the life cycle also affects the resources available to each household. Young employed adults have reasonable incomes and few responsibilities, but they have saved little with which to buy a home. Middle-aged families have higher incomes and have accumulated capital through previous house purchase, but they also have greater financial commitments. Some among the old may well have considerable savings but most have limited incomes. As a result of these different earnings/savings/outgoings ratios, they face different kinds of constraint on their housing choice.

Clearly, the housing characteristics of different age and family groups differ from society to society. None the less, important similarities emerge from studies of housing in different countries. First, private tenants seem to be drawn predominantly from young and old families throughout Western Europe and the United States (Harloe, 1985). Similar patterns are found in Latin America (Gilbert and Ward, 1985; Edwards, 1982). Findings in Bogotá and Mexico City, for example, show that housing tenure in low-income settlements is closely linked to age of household head and spouse. 'On average renters in Mexico City are 7.5 years younger than the owners, sharers 11.4 years younger; in Bogotá renters are 9.3 years younger than owners' (Gilbert and Ward, 1985: 125). Dependent on this age difference is one of household size: owner households are larger than tenant households, and tenant households larger than those of sharers. Second, single-adult households seem to be an increasingly important element in the private rental market in several developed countries. In the United States, for example, 'the ranks of renters have been increasingly dominated by households headed by women and single men. They are now almost two-thirds of all renting households, compared with less than half in 1970' (Downs, 1983: 21).

Clearly the space demands linked to rearing children are an important element influencing housing preferences. As family size increases and children grow older, the preference for more space is likely to increase. Indeed, Clark and Onaka (1983) note that the need for more space at particular points in a household's domestic cycle is associated with residential mobility and is often linked to tenure change. Childbirth, in fact, may influence tenure choice in a negative fashion even in developed countries. On the basis of Australian evidence, for example, Kendig (1984) suggests that child-rearing is a crucial stage at which households are divided permanently into renters or owners. This point has been developed by Hamnett (1986: 18) who uses childbirth as a key element to distinguish between those who enter the British public housing market and those who become home-owners. While recognising the importance of income differences: 'the great majority of recently-married council tenants had children; most recently-married owner-occupiers did not'. The principal difference between the two groups seems to lie in the presence, or absence, of the wife's income: mothers with children are less able to work than those without. This distinction is not insignificant in so far as there seems to be little mobility between the council-rental and the owner-occupier sectors. Clapham *et al.* (1987: 14) conclude that 'owner-occupation is generally accessible only at a few points in the life cycle, particularly as mortgage finance is easiest to raise when relatively young and relatively affluent'.

Whether, of course, there is such a close link in Third World cities is less well documented, but it is clear that the birth (or sometimes the death), and the eventual departure of children from the home must act as some kind of trigger in housing decisions. Whether the trigger is activated depends upon income. If a family is sufficiently poor, the addition of a further child will merely lead to more overcrowded conditions: lack of finance precludes any change of residence.

Life cycle is also linked to migration and hence to changes in residential tenure. In most societies, newcomers rent or stay with friends or kin; this is certainly the typical pattern in Latin America (Conway and Brown, 1980; Gilbert and Ward, 1982; Butterworth and Chance, 1981). On the basis of such Latin American evidence, John Turner formulated his well-known bridgeheader-consolidator model of residential movement. The first stage of this model was the move of the recent migrant into central city rental accommodation. Turner (1968) argued that residential choice was determined by a trade-off between tenure, location and shelter. As such, recent migrants preferred cheap rental accommodation in the central city to ownership in peripheral settlements. With gradual integration into the employment market, greater knowledge of the city, and growing family size, however, these priorities would change. The established migrant would now be in a position to become an owner in the urban periphery. Such a location offered space for expansion and ownership was possible through the construction of a self-help dwelling. The theory suggested that most low-income migrants would live first as renters in the inner city and later move into the peripheral 'shanty town'.

While aspects of this model have been strongly criticised, both theoretically (Burgess, 1982) and empirically (Conway and Brown, 1980), the essential fact that newcomers rent or share with kin has been supported by innumerable studies. Even if the migrants move increasingly into consolidated self-help settlements in the intermediate ring of a city, rather than into the central city (Brown, 1972; Gilbert and Ward, 1982; Conway and Brown, 1980; Vernez, 1973; Ward, 1976), the essential move into rental or some form of shared accommodation is not in question. While, of course, it is not unknown for newcomers to move directly into ownership in low-income settlements (Gilbert and Healey, 1985), it is rather unusual. The old idea that the shanty towns of Latin America are established by newly-arriving migrants has long been discredited. Certainly, the conclusion of Hoenderdos *et al.* (1983: 381), on the basis of Bolivian and Mexican data, that 'the longer one has lived in the city, the better the dream of owning a house can be realised', is generalisable to most Latin American migrants.

Of course, the formulation of the Turner model in terms of residential preference, as opposed to constraint, may be questioned. Arguably, however, it is a correct interpretation of the situation in certain West African countries, where O'Connor (1983) and Peil (1976) contend that most migrants choose to be tenants. 'Far more urban dwellers than in Latin America or south-east Asia prefer to rent accommodation, because they do not intend to stay permanently, and because entrepreneurs have made this available, sometimes at very low rents' (O'Connor, 1983: 185). In general, however, there are increasing reasons for believing that the newcomer is faced with a highly constrained range of choices on arrival in the city. Indeed, tenancy may be the longer term fate of many, including the city-born, in most Latin American and Asian cities.

## INCOME CONSTRAINTS

Underlying both the life-cycle and migrant interpretations of residential behaviour is an income-constraint explanation. For example, Doling (1976) argues that in Britain it is unclear whether the purchase of more dwelling space with increasing age and presence of children is due to changes in preferences or to changes in income. It is likely that the former provides the motivation but that the eventual outcome is a function of a household's financial situation.

What is certainly true in many developed countries is that the better off own, while the poor rent. In the United States, for example,

> in 1980, 67.6 per cent of all renters – compared with 37.1 per cent of all home owners – had household incomes below $15,000. ... Thus very few tenants are tenants by choice. They are forced by economic circumstances – and by racial discrimination in the housing market – to rent their homes.
>
> (Dreier, 1984: 261)

While there must be general agreement with that statement it is important to note certain reservations to it. First, as Downs (1983: 21) recognises, tenants in the United States are much poorer than owners 'partly because so many households switch from renting to owning as their income rises'.

Second, it is by no means certain that everyone wants or needs to own. At the very highest income levels, for example, there is some evidence of the preference for ownership becoming negative.[1]

Nevertheless, home-ownership is a widely-desired status, and there are strong signs that the ability to own fluctuates greatly from period to period even within the same country. Rising land and construction costs in many developed countries, for example, have made home-ownership highly problematic for non-owners in many developed countries during the 1980s. As Howenstine (1983: 86) argues, there may well be a growing group of 'enforced renters' who will never be able to afford to buy their own homes. 'This factor has assumed great importance in many countries, such as Australia, the Federal Republic of Germany, New Zealand and Switzerland'.

In Latin America, there is clear evidence that the tenant household is often poorer than the owner household. Evidence from Buenos Aires, Bogotá, Bucaramanga, and Mexico City (Yujnovsky, 1984: 345; Gilbert and Ward, 1985; Edwards, 1981), suggests that owners are wealthier than tenants. Despite the fact that the incomes of the principal wage-earner are often similar between tenant and owner households, family incomes are significantly different. Of course, some tenant households are more affluent than many owner families, but the general pattern is clear; estimates across the city of Bogotá as a whole showed that average incomes among owner occupiers were 50 per cent higher than those of tenants (Colombia, DANE, 1977).

While the poorest families are undoubtedly 'enforced renters', it is by no means certain that all such families will always remain tenants. Since most of the non-owners in Bogotá and Mexico City were much younger than the owners, it is likely that many will eventually accumulate enough savings to move into ownership (Hamer, 1981). In addition, as their children grow up and begin to contribute to the household budget, they will be in a better position to make the transition to ownership (Gilbert and Ward, 1985: 125).

The principal question with respect to this group of younger households is whether macroeconomic circumstances will permit them to imitate their predecessors. Given the fierce economic recession in most Latin American countries during the 1980s, and the severe falls in real incomes among the poor and even among the middle class (UNECLA, 1985), the likelihood is that the opportunities for home-ownership are declining in many countries. Indeed, where the crisis has been particularly deep, and where alternative forms of land acquisition, such as invasion, have been precluded, over-crowding has reached historically high levels. Indeed, reduced incomes and limited accommodation may have forced many into sharing with kin. In

Santiago, for example, Bähr and Mertins (1985) note that many new families are forced to share with parents and kin; even renting may be difficult. Necochea (1987) contrasts the fact that most families in Santiago had a house or plot in 1973 with the situation in 1985 when possibly 42 per cent of families were sharing homes. In the centre of Montevideo, the freeing of rent controls combined with economic recession produced much higher levels of crowding, displaced families often moving in with kin (Benton, 1987: 41).

Of course, the quality and tenure of housing do not vary with the level of income alone. They also depend to some extent on the willingness of families to spend their incomes on housing. At the margin, most families can trade off worse conditions by paying more rent. This may not apply to the very poorest – some of the pavement dwellers of India may actually be unable to pay rent – but for most families, even in less-developed countries, the share of income they are prepared to dedicate to housing will be an important variable in the residential decision (Malpezzi and Mayo, 1987a).

In the past, constraints on location and poor transport may well have increased the share of housing in the total family budget. 'In 1885 in London, over 85 per cent of the working classes spent over one-fifth of their income in rent and almost one-half paid between one-quarter and one-half. After 1885, rents rose even more steeply than before' (Wohl, 1971: 26). Recent World Bank data on income elasticities in sixteen cities in less-developed countries suggest that rent/income shares are somewhat lower than these figures (Mayo, 1985). Although in Seoul most families spend between one-third and one-half of their incomes on housing, the data for Bogotá, Manila and Cairo show that few families spend more than 20 per cent of their income on rent, and most in Manila and Cairo less than 10 per cent. As Mayo (1985) points out, however, the evidence from the sixteen cities suggests that 'the fraction of income that households allocate for housing is highly variable, depending in particular on household income and on the level of economic development'. To this, we might also add cultural factors, for there is some reason to believe that housing conditions are considered less important in some cultures than in others. Peil (1976: 135) argues that 'a case can be made that housing is a less important factor in the standard of living and self-image of a West African than of an Englishman or American'.

It is also important to remember that in a flexible housing market households may change their home regularly to match their income levels. In the Victorian English city, short tenancies meant that poorer households could move up or down the housing stock very easily. They could also change their circumstances within the house by sub-letting:

if income fell or rent increased beyond its means, a household could adjust to this change of circumstances by, say, renting one room less in a building or by sub-letting a room to another tenant. Indeed, sub-letting or 'doubling up' was a widespread reaction to economic recession.

(Kemp, 1987: 9)

The ability of families to adjust in this way, of course, is critically affected by the nature of housing supply and the contractual terms on which it is bought or rented.

## CHOICE WITHIN CONSTRAINT

Much of the housing literature falls fairly clearly into one of two groups: that emphasising some kind of household choice model and that stressing constraints facing the household. The distinction, of course, is one of approach and it is surely difficult for anyone to deny that both factors are operating. Few families can afford precisely the accommodation they want; equally, few families have no choice whatsoever over the tenure, location, size and quality of their housing. Within every city there are those who, with similar incomes, choose different kinds of tenure, different residential locations, and different sizes of dwelling; in this sense there is no denying the importance of choice. In general, however, this book concentrates more on constraints on the urban residential decision. This is because it is concerned primarily with the position of the poor in Mexican urban society, a group that is much less able than the rich to choose freely in the housing market. Rational choices are made by the poor, but only from a highly constrained range of alternatives. Despite this emphasis, however, we are fully aware that the choice made is highly variable as between groups and between cities. As this literature review has already demonstrated, there seem to be limits on most generalisations about housing behaviour. Indeed, we believe that the only satisfactory way in which to explain these variations, is by accepting that both choice and constraint are critical ingredients in the housing decision; residential decisions cannot be properly understood except in these terms.

## RESIDENTIAL TRAJECTORIES IN GUADALAJARA AND PUEBLA

The transition in Mexican cities from renting to ownership implies that most ordinary Mexicans wish to become owners. Do low-income families in Guadalajara and Puebla actually express such a marked preference for ownership? Our interviews with tenants seemed to demonstrate beyond doubt that ownership is perceived to be by far the most desirable option. Some 96 per cent of tenants interviewed in Guadalajara said that they would like to own a home; for Puebla, the figure was 93 per cent.

The preference for ownership is reflected strongly in the residential sequence figures. These underscore earlier suggestions that in Latin America 'the renting-to-ownership pattern is a one-way process' (Gilbert, 1983). In Guadalajara less than 2 per cent of tenants interviewed had previously owned a home in the city; in Puebla we found no such cases. Out of a total of 411 tenant households, therefore, we interviewed only four which had previously owned a home in the city.

The results also reveal how the move into ownership takes place. Table 6.1 shows that while a minority have always owned, the majority of owners had previously either rented or shared accommodation.[2] Roughly two out of five families have moved directly from renting into ownership. A smaller but not insignificant group had moved directly from sharing into ownership. A small group had even proceeded through the whole 'sharing-tenant-owner' transition. There were surprisingly few variations from this general pattern of movement into ownership. A few households had looked after someone else's house for a period, while others had lived in accommodation provided by their employer.

If we look at the moves of the tenants we find that most follow the 'expected' pattern. A majority have always rented a home; a significant minority had moved from sharing into renting. Few tenants have had any other kind of tenure.

What these figures fail to reveal, however, is whether or not the average tenant's seeming preference for ownership can be realised. Are tenants constrained in some way from moving into ownership? Do tenants in Guadalajara and Puebla find difficulty in moving into peripheral ownership because of their low incomes? Are there other explanations for their continuing in rental accommodation?

To clarify this point, we must necessarily turn to comparisons of income, age and household characteristics: who are the tenants and who are the owners? When the earnings of the male householder are compared across the different tenure groups, we find little evidence that tenants are poorer on average than owners. In Guadalajara there is remarkable similarity: a median of between 40,000 and 43,000 pesos per month (1985 values) for all groups.[3] In Puebla, however, there is some difference between the groups. Incomes of male householders in the older self-help settlement are the highest, with a median value of almost 47,000 pesos per month; among the other three groups, male householder incomes all fall between 34,000 and 38,000 pesos per month.

*Table 6.1* Summary of household residential histories (per cent)

| | GUADALAJARA | | | | PUEBLA | | | |
|---|---|---|---|---|---|---|---|---|
| | YSO | OSO | OST | CCT | YSO | OSO | OST | CCT |
| Always owned | 6 | 25 | – | – | 22 | 21 | – | – |
| Always rented | – | – | 70 | 78 | – | – | 67 | 69 |
| Rented then owned | 40 | 41 | – | – | 47 | 40 | – | – |
| 'Shared' then owned | 16 | 14 | – | – | 13 | 29 | – | – |
| 'Shared' then rented | – | – | 24 | 16 | – | – | 27 | 25 |
| 'Shared' then rented then owned | 13 | 4 | – | – | 14 | 4 | – | – |
| One of the above, but including another form of tenure at some point | | | | | | | | |
| – to ownership | 18 | 9 | – | – | 1 | 4 | – | – |
| - to tenancy | – | – | 1 | 3 | – | – | 3 | 5 |
| Other | | | | | | | | |
| - to ownership | 8 | 7 | – | – | 4 | 1 | – | – |
| - to tenancy | – | – | 5 | 3 | – | – | 4 | 1 |
| Total | 100 | 100 | 100 | 100 | 100 | 100 | 100 | 100 |
| Sample size | 90 | 56 | 82 | 104 | 100 | 72 | 78 | 108 |

*Source:* Household survey.

*Notes:*  YSO – Young settlement owners
OSO – Older settlement owners
OST – Older settlement tenants
CCT – Central city tenants
Figures refer to the 690 households (92 per cent of those surveyed) whose complete residential history is known. Only the *pattern* of tenure change is recorded, not the number of houses held. For example, households who have 'rented then owned' may have lived in several different rented properties; but they have not returned to renting once ownership was achieved, and they have not had any other kind of tenure.
'Shared' –shared a plot with its owner, *or* lived in an extended household without being the householders.
'Other' – for example, houses held in connection with the householder's employment, or lent to the household, or for which the household acted as caretakers.

The picture is modified, however, when we consider household incomes.[4] Table 6.2 conforms to our expectations in so far as owners in the older self-help settlements have higher incomes than the other groups. More surprising are the comparatively high incomes of some of the tenant households in each city. In Guadalajara, central city tenants earn more than the new settlement owners. In Puebla, incomes of the young settlement owners fall between those of the two groups of tenants.[5]

Clearly, differences in income alone cannot explain tenure in the two cities.

*Table 6.2* Median household income from employment (1985 pesos)

|  | GUADALAJARA | | | PUEBLA | | |
|---|---|---|---|---|---|---|
| Young settlement owners | 50,399 | [57%] | (83) | 43,195 | [70%] | (69) |
| Older settlement owners | 60,649 | [61%] | (31) | 62,770 | [59%] | (30) |
| Older settlement tenants | 52,076 | [51%] | (71) | 44,534 | [55%] | (54) |
| Central city tenants | 56,380 | [67%] | (69) | 42,847 | [63%] | (68) |

*Source:* Household survey.

*Notes:* Figures are median total monthly household incomes from employment, in 1985 pesos. The data on which the table is based refer only to those households for which the income of *every* household member was given.
The figure in square brackets is the coefficient of variation of the data (i.e. the standard deviation expressed as a percentage of the mean).
Sample size in parenthesis.

Age is also an important explanatory variable. Our early thinking inclined us to believe that owners would probably be older than tenants (Gilbert, 1983), and some of the data supports this view. Owners in the older self-help settlement were much older than the other groups. Where the findings do not fit the expected pattern is in the age of the tenants *vis-à-vis* the young settlement owners. In Guadalajara owners in the young settlement were younger than either of the tenant groups; in Puebla they were older than the older self-help settlement tenants but younger than the central-city tenants (see Table 6.3).

At first sight, therefore, these results come perilously close to suggesting that owners in the young settlements are both poorer and younger than the tenant groups – a finding which would be at odds both with the economic 'constraints' argument, according to which owners should be in a better economic position than tenants, and the household 'choice' argument, which predicts that owners should be older than tenants, because of the life-cycle arguments reviewed earlier.

*Table 6.3* Median age of householder (years)

|  | GUADALAJARA | | PUEBLA | |
|---|---|---|---|---|
|  | *Woman* | *Man* | *Woman* | *Man* |
| Young settlement owners | 32 (101) | 34 (90) | 35 (94) | 38 (95) |
| Older settlement owners | 47 (58) | 52 (56) | 49 (74) | 49 (64) |
| Older settlement tenants | 35 (95) | 34 (81) | 32 (77) | 34 (68) |
| Central city tenants | 41 (110) | 43 (96) | 36 (110) | 39 (100) |

*Source:* Household survey.

*Note:*   Sample size in parentheses.

Such a finding would only be valid, however, if the different tenure groups contain homogeneous populations. If there were wide variations between the characteristics of tenant families, for example, this would produce an inaccurate picture of the average tenant. It might conflate relatively wealthy families occupying good rental accommodation in the central city with the occupants of dilapidated *vecindades*. We decided therefore to examine the variations within tenure groups.

Our first check on variation was to examine the distribution of income within each settlement/tenure group. We removed the highest 25 per cent of household incomes in each settlement and recalculated median values for each group. The main effect of the change was to reduce the incomes of Guadalajara tenants, both in the older self-help settlement and in the central city in Guadalajara, to a level below those of owners in the new settlement.[6]

The second check on variation within groups was to examine the different kinds of accommodation occupied by sample households. Since the quality of rental housing varied considerably within the settlements studied, it was plausible that significant differences in terms of age, income and household characteristics would be revealed by dividing the tenant population according to the kind of accommodation in which they lived. The most appropriate division was by determining whether the sample population lived in a house, a flat or a *vecindad*. While such a categorisation is not wholly defensible, it does reflect differences between homes in terms of price, space and servicing. Table 6.4 shows the distribution of the households interviewed living in the different types of housing discussed.[7]

*Table 6.4* Distribution of sample population between different housing types (per cent)

|  | GUADALAJARA | | | PUEBLA | | |
|  | House | Flat | Vecindad | House | Flat | Vecindad |
| --- | --- | --- | --- | --- | --- | --- |
| Young settlement owners | 100 | – | – (102) | 100 | – | – (100) |
| Older settlement owners | 100 | – | – (64) | 100 | – | – (76) |
| Older settlement tenants | 53 | 17 | 30 (96) | 16 | 49 | 35 (80) |
| Central city tenants | 33 | 37 | 31 (120) | 1 | 43 | 56 (115) |

*Source:* Household survey.

*Notes:* A house on a shared plot is counted for owners as an individual house; for tenants, it is counted as a flat. A room in someone else's house or in a boarding house is counted as a *vecindad*.
Sample size in parenthesis.

The data in Table 6.5 suggest that there are important differences between tenants. While many tenants are poor some are not. Certainly, some flat and house-dwellers have sufficient incomes to own in peripheral settlements if they wanted to do so. On the other hand, the inhabitants of *vecindades* are generally poorer than tenants of flats or houses in terms of income, possessions and housing amenities. It is also important to note that inhabitants of *vecindades* are generally younger than the tenants of flats or houses in the same area, particularly in the case of the older self-help settlement in Puebla, and the central city tenants in Guadalajara.

Comparing Tables 6.5 and 6.6, it can be seen that tenants living in *vecindades* are generally poorer than the people who move into ownership in a new settlement on *ejido* lands. The data from Puebla are the clearest in this respect: both groups of *vecindad* dwellers are poorer in terms of income and possessions scores than the owners in El Salvador. In Guadalajara, the differences are less clear.[8] The data, then, are not wholly conclusive, but the broad picture is one of *vecindad* dwellers performing worse on income indicators and surrogate measures than either owners in the new settlements or other types of tenant. Non-*vecindad* tenants perform rather better than owners in the new settlements.

The difference between the incomes of *vecindad* tenants and owners in the new settlements is not, however, very large. One reason for this is that some of the former are about to become owners themselves. A significant minority of the tenants interviewed had already bought a plot of land, or, in a handful of cases, even purchased a house or flat. This applied to around 5 per cent of tenants in Guadalajara, but to as many as 16 per cent of tenants in Puebla.[9] These tenants had not yet moved house because they lacked the money to build an adequate dwelling or to complete an existing house. Nevertheless, some families clearly do manage to escape from the *vecindades*. This point was further demonstrated by the fact that many owners interviewed in the young settlements had previously lived in *vecindades*. However, this in turn raises the question of whether there are indeed some who are trapped in the *vecindades*.

*Table 6.5* Selected characteristics of tenants in different housing types

| Type of house | GUADALAJARA | | | PUEBLA | | |
|---|---|---|---|---|---|---|
| | House | Flat | Vecindad | House | Flat | Vecindad |
| *Older settlement tenants:* | | | | | | |
| Income (median) | 48,070 | 50,314 | 56,082 | 107,418* | 48,067 | 34,590 |
| Income (mean) | 56,227 | 51,304 | 55,532 | 104,113* | 57,709 | 36,045 |
| Possessions | 9.8 | 7.1 | 6.4 | 9.4 | 9.4 | 6.4 |
| Housing amenity | 4.6 | 4.0 | 3.4 | 5.3 | 4.6 | 3.3 |
| Age (woman) | 35 | 33 | 34 | 33 | 35 | 29 |
| Age (man) | 36 | 33 | 33 | 37 | 35 | 31 |
| Household size | 5.8 | 5.3 | 4.4 | 5.2 | 5.1 | 4.4 |
| No. of earners | 1.4 | 1.4 | 1.4 | 1.6 | 1.4 | 1.4 |
| Length of res. history | 8.5 | 6.0 | 10.0 | 9.0 | 7.5 | 7.0 |
| Rent | 6,409 | 6,810 | 3,205 | 5,699 | 6,268 | 4,701 |
| *Central city tenants:* | | | | | | |
| Income (median) | 43,236 | 64,202 | 45,808 | – | 65,126 | 37,610 |
| Income (mean) | 68,399 | 72,570 | 60,984 | – | 69,152 | 44,280 |
| Possessions | 10.7 | 10.7 | 7.0 | 11.0* | 11.6 | 6.5 |
| Housing amenity | 5.9 | 4.6 | 3.0 | 7.0* | 4.8 | 2.9 |
| Age (woman) | 48 | 38 | 35 | 27* | 37 | 34 |
| Age (man) | 45 | 41 | 37 | 27* | 40 | 39 |
| Household size | 5.2 | 5.0 | 3.7 | 4.0 | 4.7 | 5.2 |
| No. of earners | 1.8 | 1.8 | 1.2 | 2.0* | 1.6 | 1.5 |
| Length of res. history | 17.5 | 11.5 | 12.0 | 6.0* | 12.0 | 11.0 |
| Rent | 8,457 | 6,343 | 3,524 | 8,141 | 4,884 | 2,714 |

*Source:* Household survey.

*Notes:* See Table 6.6.

*Table 6.6* Selected characteristics of owners

|  | GUADALAJARA | PUEBLA |
|---|---|---|
| *Young settlement owners*: |  |  |
| Income – median | 50,399 | 43,195 |
| Income – mean | 59,418 | 52,240 |
| Possessions | 5.1 | 8.2 |
| Housing amenity | 3.8 | 4.0 |
| Age (woman) | 32 | 35 |
| Age (man) | 34 | 38 |
| Household size | 6.2 | 5.7 |
| No. of earners | 1.6 | 1.7 |
| Length of res. history | 8 | 8 |
| *Older settlement owners*: |  |  |
| Income – median | 60,649 | 62,770 |
| Income – mean | 74,791 | 72,039 |
| Possessions | 10.3 | 11.7 |
| Housing amenity | 5.2 | 6.0 |
| Age (woman) | 47 | 49 |
| Age (man) | 52 | 49 |
| Household size | 5.8 | 5.4 |
| No. of earners | 1.8 | 1.8 |
| Length of res. history | 18 | 24 |

*Source:* Household survey.

*Notes to Tables 6.5 and 6.6:*

Total household income and rent – expressed in 1985 pesos.

Possessions – a range of consumer goods have been weighted and the possessions of each household recorded in a score reflecting this weighting.

Housing amenity – refers to the number of rooms (with the exception of bathroom and kitchen, when the latter is a small unit used for no other purpose), adjusted in accordance with the sanitary facilities available to the household, in order to produce a score. One point was taken off the number of rooms score for the absence of toilet facilities; one point was added for a shared toilet, and 2 points were added for a private toilet.

Length of residential history – where known, the total number of years passed by the household in any kind of housing in the city. The household is defined with reference to its status on arrival at the present house: if a couple (whether or not later divorced or widowed etc.), the history of that couple is given. If a widow, for example, the period involved is that since the person was widowed. Previous marriages etc. are not counted. A single person constitutes a household since starting to live independently.

Statistical measure used:   income                 – median and mean
                            possessions            – mean

| Statistical measure used: | housing amenity | – mean |
|---|---|---|
| | age | – median |
| | household size | – mean |
| | no. of earners | – mean |
| | length of res. history | – median |
| | rent | – median |

In general, the median has been preferred. The mean is used only where the maximum number of units is small and therefore the median would be a rather crude measure, revealing little distinction between the different groups.
\* – 5 cases or less. Sample sizes cannot be given because of the great number of different variables involved.

One way of examining this point is by comparing the age of the *vecindad* dwellers with the age at which the owners first obtained a plot of land in the city. If the tenants are older than the owners, it suggests that the tenants face greater difficulty in obtaining plots than did the owners. If the owners are older than the tenants, then it is more plausible that the tenants will have become owners by the time they have reached the same age as the owners when the latter acquired their plots.

In order to make a meaningful comparison, however, it is necessary to remove from the calculations all householders who moved to the city as adults.[10] Table 6.7 reveals that tenants in Guadalajara are considerably older than owners were at the time of plot acquisition.[11] In Puebla, however, while the tenants in one settlement were also older than the owners had been at the time of purchase, in the other they were a little younger.

Clearly, the variation means that we cannot make a general statement about the difficulties of obtaining land on the basis of these data. What we may infer from the data, however, is that land is not easy to obtain in either city. The fact that most male owners were in their late twenties or early thirties before they obtained a self-help plot is *prima facie* evidence of the difficulties involved in mobilising the resources to buy a plot of land. If we consider the means, the age of plot acquisition is still higher. The mean figures are also interesting in so far as they are similar to the 34 years at which self-help owners in peripheral settlements in Bogotá first purchased their land (Gilbert, 1983: 466). These figures are particularly revealing when compared to the typical age of house purchase in the United Kingdom. In the latter, 53 per cent of householders in the 25-29 year age group were owner-occupiers in 1980 (Boleat, 1985). This underlines the difficulty of obtaining land in Mexican cities. If British households can occupy their own house in their late twenties, it is a sad comment that so many Mexicans can only begin building a shelter in their early to middle thirties.

*Table 6.7* Age of *vecindad* tenants compared to age of owners at the time of plot purchase (years)

|  | GUADALAJARA | | | | PUEBLA | | | |
|  | Woman | | Man | | Woman | | Man | |
|---|---|---|---|---|---|---|---|---|
| *Age of owners at time of plot purchase*: | | | | | | | | |
|  | Md | Me | Md | Me | Md | Me | Md | Me |
| Young settlement owners | 24 | 26 (38) | 26 | 28 (27) | 28 | 32 (41) | 32 | 35 (45) |
| Older settlement owners | 28 | 30 (27) | 29 | 32 (18) | 28 | 28 (24) | 32 | 33 (21) |
| *Current age of tenants*: | | | | | | | | |
| Older settlement tenants | 32 | 33 (16) | 32 | 37 (13) | 25 | 29 (19) | 29 | 33 (14) |
| Central city tenants | 33 | 43 (23) | 34 | 40 (15) | 33 | 36 (52) | 36 | 38 (44) |

*Source:* Household survey.

*Notes:* Md – median.
Me – mean.
Only natives of the city and migrants who arrived before the age of 16 are included.
Only households purchasing their plot are included; only households whose present plot is their first one are included.
Sample size in parenthesis.

## PERSISTENT TENANTS

The interviews provide some support for the idea that some tenants are poorer than owners and are therefore prevented from establishing their own home. None the less, we also found some examples of tenants who could have afforded to own but continued to live in rental accommodation. Indeed, despite saying that they would like to own a house, a substantial minority of tenants had household incomes at least as high as those of owners in newly-established self-help areas. In particular, we found some tenants living in flats and houses in the central city who were comparatively wealthy. In terms of income, personal possessions and quality of housing they were among the wealthiest of households interviewed. If they, like the rest, would prefer to own, why do they still continue to live in rental accommodation?

One explanation is that some households, while apparently earning enough money to move into ownership, have not been able to accumulate enough savings to make the transition. Our interviews with owners showed that almost all had been obliged to make a large down-payment on their plot. Outright purchases were also common, two-thirds of the households in Buenos Aires having bought land in this way. Although half of the owners in the two Puebla settlements and two-thirds of those in Agustín Yáñez had paid through monthly instalments, the down-payment often constituted a substantial proportion of the total price. A significant problem facing any family, therefore, is to obtain the deposit on a house or plot.

Our survey revealed how some households had mobilised the necessary funds to buy their first plot of land. A few household heads in formal sector employment had used their Christmas bonus as a down-payment. One in six purchasers had bought their plot with the help of a loan, and others had access to credit through an informal savings scheme with friends.[12] Others, especially migrants, had sold property; a few had won money in a lottery. Compensation from accidents, or money from gifts or inheritance had also been used, as had money brought back from work in the United States. Apart from loans, however, the most important methods of obtaining the deposit were for another member of the household (usually the woman) to take a job, or for an existing worker to take on an extra job: such strategies were reported by 13 per cent of those who had bought their property. In total, more than half of the owners purchasing their plot reported that the deposit had been obtained either through a windfall or through some special effort being made by the household.

Some young households may also be in a better position than others to accumulate savings for a deposit as a result of their relationship with families who already own their home. During the course of interviewing in Guadalajara and Puebla, a tendency for owners to be children of owners, and for tenants to be children of tenants, became apparent. This is confirmed by the data presented in Table 6.8.

*Table 6.8* Tenure of householders' parents (per cent)

| Proportion of householders' parents who were owners: | GUADALAJARA | | PUEBLA | |
|---|---|---|---|---|
| | Woman | Man | Woman | Man |
| Young settlement owners | 62 (29) | 56 (32) | 63 (30) | 61 (31) |
| Older settlement owners | 65 (23) | 54 (24) | 56 (34) | 72 (32) |
| Older settlement tenants | 29 (56) | 31 (51) | 33 (33) | 41 (34) |
| Central city tenants | 26 (50) | 27 (37) | 11 (73) | 16 (67) |

*Source:* Household survey.

*Notes:* This table applies only to householders who formed part of a couple on arrival in their present house, and who were living in the city as part of their parents' household immediately before forming that couple (these are the only cases in which information on parents' tenure is available). It refers to the parents' tenure *at that time only.* Out of the 720 female householders 328 women, and out of the 653 male householders 308 men fall within the category stated. The figures refer to the percentage of these women and men who were living with *owner* households. The remainder were living with tenant households, except for a handful who were living with households in some other type of tenure. Sample size in parenthesis.

The significance of this finding is that children of owners may be in an economically stronger position than children of tenants to become owners themselves. Some, of course, will become owners once their parents die, or may be given the house by their parents. In Guadalajara, 10 per cent of the owners interviewed in Agustín Yáñez had received a house as a gift or inheritance; the corresponding figure for Veinte de Noviembre owners in Puebla was 20 per cent. Alternatively, children may benefit from a share of the proceeds from the sale of an inherited property. Probably more important, however, is the fact that children of owners can live with parents for an extended period without paying rent. Although they may contribute something to housing costs, they are likely to pay much less than an ordinary rent. This means that they can save a higher proportion of their income and accumulate a down-payment more quickly than the average child of a tenant. To give an idea of the importance of this factor in enabling households to buy a plot, it is interesting to note that 11 per cent of owners who had *purchased* their land were previously living with their parents; the corresponding figure for current tenants was only 5 per cent.

A second factor that may help to explain why certain tenants continue to rent is that they may be less able to undertake the task of self-help construction. Single-parent households are likely to face more difficulties in building and consolidating a home than others. Elderly households and female-headed households will also find it more problematic to move into peripheral ownership. The concentration of these households among the inner-city tenants is shown in Table 6.9. Tenants in the older self-help settlements and owners in the young settlements have similar percentages of elderly households, but more of the tenants are single-parent (usually female-headed) families.[13] Central-city tenants have a greater proportion of these types of household than the new settlement owners.

*Table 6.9* Proportions of single-parent, female-headed and elderly households, by settlement and tenure groups

|  | GUADALAJARA | | | | PUEBLA | | | |
|---|---|---|---|---|---|---|---|---|
|  | A | B | C | D | A | B | C | D |
| Young settlement owners | 5 | 5 | (102) | 7 | 8 | 2 | 2 | (100) | 4 | 6 |
| Older settlement owners | 9 | 9 | (64) | 16 | 30 | 6 | 6 | (76) | 27 | 30 |
| Older settlement tenants | 8 | 10 | (96) | 6 | 9 | 10 | 11 | (80) | 3 | 6 |
| Central city tenants | 12 | 15 | (120) | 20 | 21 | 7 | 9 | (115) | 12 | 15 |

*Source:* Household survey.

*Notes:*  A – percentage of single-parent families: nuclear households, in which the householder had resident children but no resident spouse.
B – percentage of female-headed households: single-parent families in which the householder was a woman, plus a small number of households consisting of women living alone.
C – percentage of households in which female householder is aged 60 or over.
D – percentage of households in which male householder is aged 60 or over.
Sample size for A and B in parentheses.
Sample size for C and D see Table 6.3.

Other households likely to experience difficulty in building a house include those that lack the necessary skills and experience. In this context, it is interesting to compare the employment characteristics of owners and tenants in the survey settlements. Table 6.10 indicates the high percentage of construction and building trade crafts workers to be found among the owners in both cities.[14] In Guadalajara, 38 per cent of male home-owners in the new settlement were involved in the building trade compared with only 12 per cent of all the tenants surveyed. In Puebla, the difference was only slightly less marked: 26 per cent of new settlement owners worked in construction compared with only 12 per cent of tenants. In hindsight, the discovery that new settlements are full of households with experience in construction work is hardly surprising. What it strongly emphasises, however, is that some households find self-help construction to be a much easier task than others.

A third explanation of 'persistent tenancy' is that the advantages of ownership in a peripheral settlement may well be less for some households than for others. Clearly, families differ considerably in terms of their internal characteristics. Table 6.5, for example, showed that owners in the new settlements had larger households than did tenants, and especially *vecindad* tenants. In Guadalajara, Buenos Aires households had an average of 6.2 members compared to only 4.1 among those living in *vecindades*. In Puebla, the difference was less marked: 5.7 compared to 4.9.[15] This difference in household size may well be related to the additional space that is available in a peripheral settlement. Whereas smaller families may be prepared to continue renting a room, larger families may find it more difficult. This is particularly likely given the strong preference that landlords express for small families and especially for families without children (see Chapter Eight). Some kind of push factor is certainly operating on the larger households.[16]

A fourth factor that may help to explain why some tenants continue to rent is that they may be more reluctant to face the difficulties and rigours involved in building their own home. Self-help construction, however much the family participates directly in the construction process, is clearly a major burden in terms of time and inconvenience. In addition, it involves living in a poorly-serviced settlement for a number of years. The thought of living in a settlement such as Buenos Aires, where most households lack piped water, sewerage and even electricity, may well deter some households from moving into ownership. For the more affluent tenant families who are living in reasonable accommodation in the centre of the city, this factor may be critical. They may well choose to rent until they have accumulated enough savings to buy something better. If they wait, they may acquire a plot in an established settlement, or even buy a complete or semi-finished house. In fact, the survey identified a number of families who were doing just that. In addition to those tenants who had already purchased a property, around two-fifths of the tenants in Guadalajara, and one-fifth in Puebla, said they were looking for something to buy. In both Analco and Central Camionera, around two-thirds of those still looking

*Table 6.10* Employment characteristics of male householder (per cent)

| | GUADALAJARA | | | | PUEBLA | | | |
|---|---|---|---|---|---|---|---|---|
| | YSO | OSO | OST | CCT | YSO | OSO | OST | CCT |
| *Jobs:*[+] | | | | | | | | |
| Trader | 6 | 13 | 12 | 13 | 12 | 16 | 13 | 21 |
| Construction worker | 29 | 21 | 9 | 6 | 18 | * | * | * |
| Driver | * | 9 | 11 | 7 | 14 | 19 | 8 | 12 |
| Factory worker | 17 | 9 | 9 | 12 | 9 | 9 | 6 | * |
| Building trade crafts | 9 | 6 | 9 | * | 8 | * | 8 | 10 |
| Office worker | * | 9 | * | 15 | * | 9 | 11 | * |
| Mechanic | * | 6 | 7 | 7 | 7 | * | 9 | * |
| Police/security | * | * | * | * | 9 | * | * | 11 |
| Shoe industry worker | * | * | 17 | * | * | * | * | * |
| Textiles industry worker | * | * | * | * | 6 | 11 | * | * |
| Car industry worker | * | * | * | * | * | * | 6 | * |
| Personal/domestic services | * | 6 | * | * | * | * | * | * |
| Porter | 7 | * | * | * | * | * | * | * |
| Musician | * | * | * | 6 | * | * | * | * |
| Other | 10 | 15 | 12 | 22 | 9 | 16 | 27 | 20 |
| Number of replies | 87 | 47 | 81 | 85 | 90 | 57 | 64 | 91 |

*Source:* Household survey.

*Notes:*   YSO – Young settlement owners
          OSO – Older settlement owners
          OST – Older settlement tenants
          CCT – Central city tenants
          * – 5% or less.
          [+] – For notes on job types, see note 14.

said they wanted to purchase a house or flat rather than a plot of land. While this was not the case for other aspirant owners, and while the hope to buy an existing house or flat may be highly unrealistic for many families, it does explain why some less poor tenants have not moved into peripheral ownership.

For the poor tenant living in a crowded, poorly serviced *vecindad*, the choice may be less clear cut. The choice between ownership and tenancy also involves a choice between life-styles. In several respects, life is very different in peripheral and more centrally located areas. Each offers advantages and disadvantages which may be crucial for particular families.

*Vecindades*, for example, suffer from severe overcrowding, small and often windowless rooms, deteriorating physical conditions and limited services; traffic noise and pollution are both likely to be a nuisance for inhabitants. On the other hand, they offer a more central location and, even if the services are poor, the tenants normally have access to water and electricity. In contrast, self-help settlements offer space but fewer services. They may be dusty in the dry season, muddy when it rains. They offer possibilities for investment and consolidation but they are located far from the central city. Individual households have different propensities to accept these different kinds of conditions. Location, in particular, may be crucial. For those working in the central area, relocation to the periphery of the city may involve a long daily journey. Irregular hours of work or the need to return home frequently may preclude the choice of peripheral settlement.

It is likely that in choosing between these different life-styles, past experience may become a critical factor in selection. Two aspects of past experience are worthy of mention here: a family history of ownership and location of previous residence. Our survey reveals two interesting differences between owners in the new settlements and the average tenant. The first is that more owners seem to have been brought up in families which owned property (Table 6.8); the second, that owners are more likely than tenants to have been born outside the city and, in particular, to have been born in rural areas (Table 6.11).[17]

The significance of past ownership is that children of owners may well be prepared to make greater sacrifices than tenants in the cause of becoming a home-owner. Certainly, such an attitude emerged from some extended interviews carried out to complement the questionnaire survey. One man in a peripheral settlement in Puebla, for example, described the lengths to which he and his wife had gone to build a home of their own. He stressed the 'sacrifices' that 'people like us' are prepared to make in order to leave something for the children. He also criticised the tenants living in the centre of the city who, in his opinion, would not make an effort to save money, thinking only of today and spending their money on entertainment or consumer goods.

Similarly, the fact that more owners than tenants originated in rural areas may be linked to their greater tolerance and experience of rudimentary housing, services and infrastructure.[18] Rural housing in Mexico is typically far more flimsy and poorly serviced than urban housing. Migrants originating in rural areas are more likely to be tolerant of conditions in peripheral settlements than are those who have always benefited from solid walls and piped water. Peripheral settlements also provide them with a more familiar environment: they have room to raise a few turkeys or rabbits, and space to grow a few vegetables. These may be used as a source

*Table 6.11*   Householder migration characteristics (per cent)

| | GUADALAJARA | | | PUEBLA | | |
|---|---|---|---|---|---|---|
| | Natives | Rural migrants | | Natives | Rural migrants | |
| **WOMAN** | | | | | | |
| Young settlement owners | 15 | 78 | (100) | 37 | 71 | (94) |
| Older settlement owners | 28 | 67 | (58) | 44 | 57 | (75) |
| Older settlement tenants | 38 | 62 | (94) | 47 | 51 | (77) |
| Central city tenants | 39 | 55 | (108) | 66 | 54 | (110) |
| **MAN** | | | | | | |
| Young settlement owners | 14 | 76 | (88) | 36 | 75 | (95) |
| Older settlement owners | 23 | 61 | (56) | 40 | 59 | (65) |
| Older settlement tenants | 39 | 60 | (79) | 46 | 51 | (68) |
| Central city tenants | 33 | 58 | (95) | 68 | 50 | (101) |

*Source:*  Household survey.

*Notes:*   'Natives'        –  proportion of all householders who were born in the city.
                'Rural migrants' –  proportion of all migrant householders who were born in a
                                           rural area.
                Sample size in parenthesis.

of extra income, or as a contribution to the family diet at times of extra economic hardship. Overcrowded *vecindades*, where such options are not open to the tenants, may seem especially unpleasant to migrants from rural areas.[19]

A fifth explanation of 'persistent tenancy' is that there are usually good reasons for remaining in an area where the household has been living. In particular, people may wish to retain close contact with relatives living in the area. Inertia is also encouraged by the disruption that a move would bring to children's schooling, to friendship networks, to leisure activities, or to the journey to work. On balance, therefore, some families may decide to stay in the area where they are living rather than move to the periphery of the city, even though that offers the only chance for ownership. Certainly, our data reveal that many households were subject to inertia in their choice of settlement. Table 6.12 records the proportion of households with a previous residence in the city, who had decided to stay in the same area when they last moved.[20] It is clear that tenants, and in particular those living in central areas, tend to remain in the same area when they move house. Even those who had moved to a different settlement had often moved only short distances. While this is most evident among the tenant families, it is also clear that a high proportion of new owners in Buenos Aires, for example, had originated in nearby settlements.[21]

*Table 6.12*   Households whose previous house was in the same area as their present house (per cent)

|  | GUADALAJARA | | PUEBLA | |
| --- | --- | --- | --- | --- |
| Young settlement owners | 5 | (93) | 3 | (78) |
| Older settlement owners | 27 | (44) | 47 | (49) |
| Older settlement tenants | 38 | (68) | 45 | (55) |
| Central city tenants | 67 | (93) | 50 | (72) |

*Source:*   Household survey.

*Notes:*   Figures give the percentage of those households with a previous house in the city, whose last place of residence had been another house in the same settlement. Sample size in parenthesis.

Table 6.13 confirms that many tenants had chosen their current area because kin or friends lived near by, because it was close to their place of work, or simply because they 'liked' the area (perhaps for these same reasons). Different factors, however, had a different weighting in different settlements. Among the tenants in the centre of Guadalajara, proximity to work was not an insignificant factor. This was undoubtedly linked to the number of drivers and mechanics living in the area; their 'office' was either the coach station after which the area is named, or one of the numerous mechanics' workshops located in the same area. Similarly, there were a number of musicians and waiters in the area who worked in the red-light district near by.

Under these circumstances, there may well be a genuine conflict between the tenants' desire to own and the inconvenience of living in a different location. If ownership were possible in the central city at a price they could afford, they would doubtless prefer it to tenancy. However, such an option is not on offer; the choice is between central tenancy and peripheral ownership. Given this choice, some will choose the former.

Finally, some tenants who can afford peripheral ownership may remain in rental accommodation because rents are low. This is certainly the situation of those families living in Mexico City covered by the rent freezes of the 1940s (Aaron, 1966). While the amount of such accommodation has

*Table 6.13* Reasons why people chose to live in their settlement

| | GUADALAJARA | | | | PUEBLA | | | |
|---|---|---|---|---|---|---|---|---|
| | *YSO* | *OSO* | *OST* | *CCT* | *YSO* | *OSO* | *OST* | *CCT* |
| The only area where they knew there were houses/plots for sale/rent – or chance | 28 | 13 | 43 | 30 | 20 | 38 | 34 | 24 |
| House/plots were cheap | 49 | 49 | 9 | 5 | 35 | 20 | 5 | 8 |
| Liked/were familiar with area | 14 | 17 | 14 | 16 | 31 | 22 | 14 | 32 |
| To be near kin/friends | 7 | 21 | 21 | 10 | 12 | 8 | 18 | 12 |
| To be near work | 3 | 4 | 2 | 16 | 6 | 18 | 9 | 8 |
| Other locational advantages (near centre, amenities etc.) | – | – | 1 | 12 | – | – | 3 | 7 |
| Other (including reasons exclusive to tenants) | 6 | 6 | 10 | 11 | 5 | 6 | 18 | 8 |
| Sample size | 88 | 53 | 91 | 115 | 83 | 50 | 79 | 115 |

*Source:* Household survey.

*Note:* YSO – Young settlement owners
OSO – Older settlement owners
OST – Older settlement tenants
CCT – Central city tenants
Figures for owners do not sum to 100 per cent, because respondents often mentioned more than one factor influencing the choice of area. Only one answer was recorded for tenants.

always been negligible in Guadalajara and Puebla (Mexico, SHCP, 1964), the rent-income ratios in the two cities, and indeed in Mexican cities in general, still appear to be low compared to the ratios typical of certain other Latin American cities (Malpezzi and Mayo, 1987a). This issue is explored more fully in Chapter Seven, but the finding suggests that some families are trading off poor accommodation for low rents. Certainly, our survey discovered a number of more affluent households in central areas paying less than 10 per cent of their monthly income in rent. Given the advantages to them of living in the central area and their possible aversion to engaging in self-help construction, renting at such low rents may be a satisfactory housing option. The most favourable interpretation of this situation is that in Guadalajara and Puebla there is a genuine tenure choice for some families, since relatively low rents and land costs permit either peripheral ownership or central renting.

CONCLUSION

The results from Guadalajara and Puebla clearly support the preference for home-ownership found among most urban Mexicans. Most families expressed a strong desire to become home-owners. The results also show that few households ever give up ownership to return to shared or rented accommodation. It is clear that there is a generalised residential transition from sharing and/or renting into ownership.

The results also suggest that there are important differences in terms of the income, age and household characteristics between owners and tenants. However, the data fail to support the idea that all tenants are poorer than all owners. Within the tenant and owner categories there is a great deal of variation. Among the tenants, for example, some are quite affluent, many very poor. However, it is possible to compare sub-groups of tenants with sub-groups of owners to shed light on whether the transition to ownership is feasible. It does seem that one important group of tenants is generally poorer than most owners. Our data show that the occupants of rooms in *vecindades* are both poorer and older than other households newly acquiring plots on the edge of the city. This finding generally supports the idea that many *vecindad* dwellers are constrained by income from ownership. At the same time, our data also show that some of the *vecindad* dwellers are themselves moving into peripheral ownership. The transition may not be easy but some are still managing to achieve their goal of ownership.

It is quite clear from the data, therefore, that a simple constraint model does not adequately explain residential decision-making in the two cities. With similar incomes some households choose to move into peripheral ownership, whereas others choose to rent. Given a variety of housing alternatives, households with different socio-economic characteristics may make different choices. While none of the choices facing the poor are ideal, for some households the bundle of advantages linked to a particular location determine the tenure choice. Rather than tenure always determining location, a particularly favourable location sometimes determines tenure by discouraging the move into peripheral ownership. Notwithstanding the general wish of virtually all tenants to move eventually into ownership, locational factors may lead to their remaining in rental accommodation.

This point recommends that we do not look at tenure alone but that we examine the different sets of alternatives facing low-income families. There is not only a choice between owning and renting; more accurately there is a choice between the *kind* of renting and the *kind* of ownership. While ownership is the longer-term preference and the possible outcome for many current tenants, some tenants refuse to contemplate the peripheral ownership option given the disadvantages that this would create for their style of life. This choice appears to be particularly significant for those living in central locations.

Choice of this kind, however, may be peculiar to these two cities. The array of choice between different forms of owning and renting may be less balanced in other Mexican cities, and certainly in many other Latin American cities where access to land is particularly difficult. It certainly does not imply either that access to ownership is easy in the two cities, or that housing conditions are satisfactory (see Chapter Five). Nor does it mean that the situation cannot change. Rapid rent rises or speculative increases in land prices may change the balance of forces; so too might changes in transport systems making peripheral locations more accessible to major centres of employment. Should the economic environment in either city change then either the constraint or the choice model may become more influential. Under current economic conditions, it seems more likely that it will be the constraint model that will first become the more important; falling real incomes place a critical constraint on the opportunities for effective self-help construction.

## APPENDIX: RESIDENTIAL HISTORIES

The following case studies provide a more qualitative picture of residential choice and constraint by focusing on the experiences of individual families. They are drawn from a number of in-depth interviews conducted during the study and also on interviews made over a number of years during the mid-1980s by ITESO (*Instituto Tecnológico y de Estudios Superiores de Occidente*) students in Guadalajara.[22] The six cases are not intended to be 'typical' of all tenants or owners; they are merely a way of 'bringing to life' some of the arguments made in the rest of the chapter.

### Carlos and Martha Ramírez

Carlos Ramírez and his wife Martha had recently become owner-occupiers buying a plot in a settlement formed illegally on *ejido* lands on the eastern periphery of Puebla.[23] San Antonio Abad is a recent settlement, founded in 1983, and lies about 6 kilometres east of the city centre.

Martha was born in 1949 in a small town about 70 kilometres south-west of Puebla. Her husband is the same age and was born in a small rural community about 70 kilometres east of Puebla. They met in his village, where Martha was working as a primary school teacher, and married in 1975. They continued to live in the village, in the schoolhouse provided rent free by the local community. They had two children, a girl and then a boy. Unfortunately, the little boy developed bronchitis and both children were sent, on doctors' orders, to live with Martha's parents. In order to try to reunite the family, the couple moved to Puebla in late 1979. Martha was never able to find a job in Puebla and was forced to keep moving from school to school; her current school is 30 kilometres from the city. Carlos has been looking for a better job ever since moving to Puebla, but so far without success. He works as a bricklayer, moving from job to job and spending two to three months on each building site. He is from a peasant family but worked for some time in a car factory in the state of Hidalgo until he was made redundant. Martha's salary is slightly higher than the minimum wage and rather more than what Carlos earns.

When the couple arrived in Puebla, they decided to look for a house themselves, rather than relying on friends. They found a place to rent by looking in the newspapers; a mistake, they later realised, because it's cheaper to contact private owners directly. The first home they found was in Veinte de Noviembre; it was a three-room flat, with its own toilet and washing facilities; five other families occupied apartments on the same plot. They shared it with Carlos' sister and stayed there for a year, leaving when their contract expired and the owner wanted to double the rent. They moved to the first alternative they could find, a tiny room a few blocks away. Because of its size they only stayed for a few days before moving to a house in *colonia* Joaquín Colombres, an older subdivision near to San Antonio Abad. The house had three rooms but no services except electricity. They stayed here for a month, then left because there was no lock on the back door and they were robbed. In desperation, they moved in with Carlos' sister and her new husband who were living in a small flat near by. When they could, they paid their share of the rent and they stayed here for almost a year. They moved because the rent was increased substantially and because a relative who had been storing some of their possessions needed the space. Their next home, which they found through a friend, was in an older *ejido* settlement adjacent to El Salvador. It was a rented house with three rooms, and limited toilet facilities, on a plot shared with two married children of the owner. Carlos and Martha managed to stay here for three years, during which time their children at last came to live with them. Eventually, however, the owner evicted them, when a daughter separated from her husband and needed somewhere to live in a hurry. Their next move was to a small three-roomed house in a nearby settlement, which they rented for the next six months. In 1985, they moved into their own home in San Antonio Abad. They had used their Christmas bonus as the deposit on the $200 \text{ m}^2$ plot and had been paying monthly instalments for the past eighteen months. They were a little concerned about the legality of the transaction, but the *ejidatario* assured them that it was a 'serious' sale and that Agrarian Reform Ministry officials had given their tacit approval to the clandestine development. The couple were also reassured by the fact that they had connections with the vendor through relatives. In addition, Martha's mother also bought a plot, just opposite theirs, and came to live in Puebla.

Carlos and Martha started to build in August 1985, when they finished paying for the plot and had some money with which they could buy some materials. They put up a temporary three-room home which took them about one week, with help from their neighbours and members of a local Catholic self-help group (*comunidad de base*). At the time of interview, in 1986, the household was laying the foundations for a permanent structure; they had already built a concrete water tank, which they shared with their neighbours, to store the water delivered by tanker. They initially stole electricity and have since had it officially installed.

Martha and Carlos were very clear about why they preferred to be owners. First, they wanted something that they could leave for their children, somewhere where the children 'can live peacefully'. Second, they didn't want to pay rent and have the threat of eviction hanging over their heads. Finally, they had experienced some problems with tenant neighbours; it was easier to mind your own business in your own house.

### Felipe and Esther Romero

The Romero family are also recent owner-occupiers living in San Antonio Abad. Esther and Felipe both come from a small village in Oaxaca; at the time of interview, she was 28 and he was 32. Esther first arrived in Puebla when she was 12 years old, having been sent by her family to work as a nanny. She kept in touch with her home village and went out with Felipe during return visits. When she was sixteen she became pregnant and the couple decided to get married. At the time of interview, they had five children: three boys of 11, 7 and 4, and two girls, one aged 9 and one a baby.

The couple lived with Felipe's parents for a year, working their land near the village. They moved to Puebla because Felipe's mother sent them to the city to work because she needed some money. At first, Felipe worked as a bricklayer's assistant, eventually becoming a qualified bricklayer.

The couple first rented a room in a *vecindad* with eight other families. They found the room through a woman who Esther had got to know while she was working in the city as a servant; this woman was the *vecindad*'s caretaker. The room was barely large enough for the family, and they had to share the services with the other residents. Esther and Felipe stayed there for only three months. They left partly because Esther had an argument with another resident, and partly because they decided to move to Mexico City to live with Felipe's sister. The couple stayed in Mexico City for three years, renting five different houses during that time. They eventually returned to Puebla because Esther had a miscarriage and being very ill needed someone to look after her. They moved in with Felipe's sister, who had earlier returned to Puebla, in a house in *colonia* Revolución. The house lacked piped water and drainage; the only service available was electricity. There was not much room in this house, because they shared it with Esther's parents, her brother, his wife and their three children. Later, the owner also decided to live in the same property.

Four years after moving in, they found the money to put down on the plot in San Antonio. In September 1983 they moved to the plot, living in a one-room shack. It was cramped and dingy, with an earth floor, but they were forced to move because they couldn't afford to pay rent as well as the monthly instalments on the plot. When interviewed, they were behind with their payments. The *ejidatario* had recently visited them to say that he now wanted more than double the price they had originally agreed three years earlier; but they weren't too worried by this threat, although they don't have any papers for the plot other than receipts for their monthly payments, as the *ejidatario* was drunk at the time.

Originally, it was Esther who had convinced her husband that they should buy a plot. She was worried because Felipe was an alcoholic and spent far too much on drink. Even though she took in washing and ironing, they often got behind with the rent. More than that, however, she felt that 'you *have* to buy, for the kids' sake, even if it's only a little bit of land'. She had managed to save half the deposit by hiding money from her husband; she got a loan for the other half. The family were very poor, having no consumer goods except a radio. When Esther had had her latest baby, some of the neighbours gave her baby clothes and a blanket; she said that none of her other children had been so lucky.

## Manuel and Guadalupe Rivera

The Rivera household own a small house in Santa Margarita, an unauthorised private subdivision in the north-west of Guadalajara which dates from the late 1960s. It is a household of six people: the couple plus three daughters of 13, 11, and 10 years of age, and a 2-year-old son.

Manuel and Guadalupe both come from large peasant families and were born near Tepatitlán, 60 kilometres east of Guadalajara. Manuel was born in 1937, and Guadalupe is three years younger. Manuel had six years' primary school education and Guadalupe, four. Like many people from this region, Manuel worked for some time in the United States. He also worked for a short time in Guadalajara before returning to agriculture in his village.

When they got married, the couple were financially secure thanks to Manuel's savings from his work in the United States. They moved to Zapopan where Manuel had secured a job as caretaker for a group of houses. The job had the major advantage of providing them with rent-free accommodation. They lived here for six months, during which time their first daughter was born. When Manuel was given another job, as a gardener, by his employer, the family moved to a house in an illegal subdivision not far from the centre of Zapopan. The house was passed on to them by the previous gardener, but as it did not come with the job they had to pay rent. The house had two rooms, but no services except electricity. The couple had no contract and no security of tenure, but they stayed here for two years, before moving to another part of Zapopan, closer to Manuel's work. Their new home was larger and better constructed than the old one and they soon got on well with their neighbours. During their time in this house, their second daughter was born and Manuel began driving a lorry. They stayed here for two years before they were evicted because the owner claimed that he was going to carry out repairs to the house. In the meantime, however, they had put down a deposit on a plot of land in Santa Margarita. When they were evicted they moved there into the small shack that they had built. They could not afford to improve their home and continued living in the one room for the next three years. Fortunately, the area already had water, drainage and electricity, and they did not have to wait long before the streets were paved. Eventually, however, they began to extend their house: they added a room built of more permanent materials, and then pulled down the original *adobe* structure to replace it with brick walls, and completed the rest of the first floor. In the early 1980s, they built an extra room on the second floor.

One reason why the family was not able to build a better house very quickly was that Manuel had been given the opportunity to buy the lorry he drove, which he managed to do with the help of a loan. As an independent lorry-driver, his income began to improve considerably, and the family also acquired a number of consumer goods. The births of their third and fourth children completed the family; Guadalupe has never sought paid employment.

Although Santa Margarita is a long way from the centre of Guadalajara, its proximity to the amenities of the centre of Zapopan means that the family can satisfy most of its necessities in the vicinity.

## José and Isabel Zaragoza

José and Isabel and eight of their ten children rent a house in Jardines de Guadalupe, an unauthorised but now consolidated subdivision about 2 kilometres north-west of Agustín Yáñez.

José and Isabel were 42 and 43 years old, respectively, at the time of interview. They were both from Fresnillo, a mining town in the state of Zacatecas, to the north of Jalisco. After leaving school without finishing his primary education, José worked as an agricultural labourer but, every rainy season, he spent a few months working as a bricklayer in Guadalajara. Isabel, who was living with him, accompanied him on these visits. They rented a house near to wherever he was working in the city. They decided to get married when their first child was born and decided to move permanently to Guadalajara. They had no difficulty in finding somewhere to live, and moved into a cheap room in a central *vecindad*. They shared the room with relatives and most of the other rooms in the building were also rented by members of their family. After a year in these cramped conditions, they moved, because a close friend lent them his house while he was working away from Guadalajara. The house was located in an older self-help settlement about 20 minutes by bus from the centre of the city. Although the streets were still unpaved, all the main services had been installed. The family remained here until their friend returned to the city, almost a year later. They then rented a room in a *vecindad* somewhat nearer the centre. Conditions were very cramped because they now had three children and by the time they left four years later the family had grown to seven. The main disadvantage of this rented house was the lack of space. After four years, all the tenants were evicted simultaneously and this family moved to a nearby flat, a major improvement because they now had three rooms plus a kitchen. Space was still at a premium, however, because the family kept growing: during the nine years that they lived in this flat, they had five more children. Eventually, they were again evicted, because the property was flooded. José and Isabel decided to move to a rented room in Jardines de Guadalupe, somewhat further from the city centre. Money was very short because the whole family of ten depended on José's earnings as a bricklayer. The room was totally inadequate for the family. They would like to move but could not afford better accommodation; they had been living in these conditions for four years when interviewed. Ideally, they would like to own their home; but they simply could not afford to do so. Two of the older children now had jobs, as a labourer and a shop assistant, but there were still six children under fifteen.

## Adolfo Méndez

Adolfo Méndez lives in a rental property with seventeen other tenant families in the centre of Puebla. He was 77 when interviewed. Although Adolfo now lives alone, his daughter Celia lives near by with her son, and Adolfo spent most of his days with them. Celia, in her thirties, is divorced and did not have regular employment, so Adolfo gave her money from his pension; in return, she prepared his meals. He had not moved in with her, however, because he liked to have his own place to go back to, particularly when he'd been out drinking. He had lived in the same small room for forty years, longer than most of the other tenants. Three other children lived with their families, in Mexico City and Puebla.

Adolfo was born in Puebla and had always lived in Analco, mostly in *vecindades*. He was married twice, and both his wives came from the same area. After marrying his second wife in 1935, Adolfo found a home for them in an area just north of Analco. It was a single room in a *vecindad*, and they lived here for two years, before moving so that Adolfo could be closer to his work in a textile factory in the north-west of the city. Their new home was also one room in a *vecindad*. They spent several years here before moving back to the same type of accommodation near Analco. Adolfo could no longer remember much about that move, but he clearly recalled their move to his present home. The couple had heard that some new rental rooms were being built; unusually, each family was going to have its own services (toilet, washstand etc.). They had a look at the new property and thought that 'the rooms looked nice', so Adolfo went to see the caretaker and paid a deposit. The family moved in two months later. In the 1940s, Adolfo thought about buying a plot, but the idea never really appealed. His wife was very enthusiastic about the idea, and tried to get him to save for a deposit, but Adolfo considered ownership 'a headache', and used what spare money he had to buy consumer goods, such as a record-player and TV, on credit. Years later, when his wife died, he regretted his indecision. By then, however, all his children were married, and there hardly seemed any point in building a house just for himself, even if he could have afforded to do so.

Adolfo got on well with his former landlord. When the textile factory where he was working closed, the family were unable to pay the rent for six months. Adolfo went to see the landlord who readily agreed to let him pay the arrears as soon as his circumstances improved. After the old landlord died, subsequent owners were not so accommodating, but Adolfo was usually able to pay the rent while working in the textile mills. He lost his last job when the factory closed in 1966; although he continued to look for work, he was told that he was too old. Since then, he has lived and paid the rent from his small pension and whatever his children are able to give him.

## Ignacio and María del Carmen Ortega

Ignacio Ortega and his wife María del Carmen live in a *vecindad* in central Puebla. They were both born in the city; at the time of interview, Ignacio was 50 and María del Carmen was in her early thirties. They had two daughters, aged 17 and 9, and three sons, aged 15, 14 and 11. None of the children was in employment and all still lived at home. Ignacio worked in a textile factory until it closed, and is now a casual labourer, working as a bricklayer, carpenter or whatever, in the construction sector. He earns less than the minimum salary. His wife sells shoes part-time in central Puebla; the income 'helps' the family budget.

The couple met in the *vecindad* in Analco where their families lived. Their first home together was a large room in the same *vecindad*. The rent was very cheap, because the *vecindad* regularly flooded: up to 3 feet of water during periods of heavy rain. After five years, they decided to look for somewhere else to live. They moved to their present home in Analco, where they rented a single, windowless room, with a little kitchen and toilet outside. The property has eighteen virtually identical units arranged on either side of a central corridor. Unusually, Ignacio and María del Carmen had built an additional room on the roof, even though the roof is old and dangerous. They knew that conditions in the *vecindad* were bad before they moved in; in fact, the administrator warned them that the walls were not whitewashed, and there were cracks in the walls and ceiling. The rent was considerably higher than in their previous house. When it kept rising they decided to look for somewhere else. Unfortunately, they could only find more expensive accommodation, in even worse condition. In the end, they decided not to move after all, and had been living here for nine years when interviewed.

In 1984 María del Carmen and Ignacio decided to become home-owners. Although Ignacio's sister had her own housing plot, they didn't want to share it, and all of María del Carmen's brothers and sisters also rented. The family therefore bought a plot of land in an *ejido* and paid for it over a period of eighteen months, but did not have enough money to buy building materials. Eventually, however, they would have moved to their plot had circumstances in the *vecindad* not changed. In early 1986, a government agency offered the tenants a loan to buy their present homes. Ignacio and María del Carmen had soon decided to do so, preferring that option to building on their own plot of land. Nevertheless, they have retained the plot as security in case anything goes wrong with the government-sponsored purchase scheme.

# 7  Landlords and the economics of landlordism

In few countries of the world is the general image of the landlord a particularly favourable one. In places this negative picture goes back some years. In Britain, the United States and France, the nineteenth-century city was often described as being controlled by wicked landlords owning rows of terraced houses or huge tenement blocks. Such landlords were hardly renowned for their gentle treatment of tenants and many a story appears in fiction of landlords throwing entire families of tenants onto the street. Clearly, large-scale landlords did exist and some, no doubt, deserved their bad reputation. And yet, recent accounts suggest that the large landlord was hardly the dominant figure in most nineteenth-century cities. Landlords may or may not have been kind, but on the whole they did not own a large amount of property. 'In mid-Victorian Liverpool, for example, landlords of working-class property held on average between six and eight dwellings each ...' (Englander, 1983: 52). In the United States and five Western European countries, before 1918, 'it was the small local investor who provided most rented housing. At a time when their possibilities for investment were limited and risky, investment in housing was probably the main outlet for such capital' (Harloe, 1985: 13). Many of these landlords were 'working-class made good'; they were traders, publicans, builders or shopkeepers who had invested their savings in property.

The small landlord has continued to be the predominant figure in most countries retaining a private rental sector. In the United States, for example, Downs (1983: 2) notes how small-scale operators dominate: 'About 60 per cent of all rented units are in structures with fewer than five units and a third of these are single family homes'. And, in Britain, even if some large property-owners survive, the majority of landlords hold less than five properties (MacLennan, 1982: 212) . However, there does appear to be a difference in the pattern of ownership when the worst rental housing is compared with the rest. In the United States, the slum areas 'have the highest concentration of major owners. The owners tend to specialise in slum properties, owning no other type' (Sternlieb, 1969: xiii).

In less-developed countries, the literature gives a less clear picture of the landlord. In part this is because we know very little about the landlord or about property ownership; in part, because landlords seem to be rather diverse, even within the same city. In Latin America, large-scale landlords certainly exist, although it seems as if they were much more active during the nineteenth and early twentieth centuries. In Argentina, urban landlords frequently owned numerous *conventillos* producing a good income; they were 'among the richest and most respected men of Buenos Aires' (*La Prensa*, cited in Scobie, 1974: 154). Similarly, in Rosario, renting property was 'a popular form of business among the most influential ... of "decent" society' (Armus and Hardoy, 1984: 40). In Chile, the respectability of the landlord was no doubt enhanced by the fact that in Santiago the priesthood was a substantial property owner (Violich, 1944: 73). It was also helped in places by the fact that the middle classes continued to live in rented accommodation. In Mexico, rental housing remained the principal form of middle-class housing until the early 1950s; an older propertied class remained intact despite the Revolution and was joined by a number of revolutionary leaders who had invested in urban property in the 1920s and 1930s (Perló, 1981: 13 and 44).

Gradually, however, the profitability of this type of investment was falling and capital began to seek out alternative areas of investment. The rent controls introduced during the Second World War were a further discouragement. While we have too little information to be absolutely certain, it seems as if few among the elite now invest in rental housing in Latin American cities. There has been little interest in investing in new rental property for many years.

In some other parts of the Third World, however, rental property still constitutes a lucrative business. In Nairobi, Amis (1982) notes that it is powerful political personalities who control the increasing network of rental housing in the squatter settlements, and in Lagos there are many cases of politically powerful owners letting rooms in blocks with up to 150 residents (Barnes, 1982). However, even in these two countries the large-scale landlord may be atypical. In Lagos there are also many small-scale landlords and access to property is eased by the existence of a communal land-tenure system. Similarly, in Nairobi, 'the majority of houses are owned by clerks, manual workers, and small-scale entrepreneurs' (Peil and Sada, 1984: 298). General reviews of the housing situation in Africa tend to emphasise the ubiquity of the small landlord: 'Some landlords are wealthy, but most own only one or two buildings, and some are just as poor as their tenants' (O'Connor, 1983: 191). In so far as the Nairobi picture differs from that elsewhere in Africa it is because the political influence of the landlords has discouraged the state from interfering; in other cities, such as Lusaka and Dar es Salaam, the state has taken action against the large landlord (Amis, 1987; Stren, 1982).

In Latin America, a similar picture of small-scale landlordism seems to be emerging. In certain Colombian cities, landlords are mainly owner-occupiers and even absentee landlords rarely own more than two or three properties (Edwards, 1982; Gilbert, 1983). Not infrequently the principal form of accommodation is provided in consolidating self-help settlements where resident owners rent out individual rooms. Petty landlordism seems to be the rule and is part of the process by which more established owners try to increase their incomes. In Bogotá,

> most landlords are themselves poor and are renting one or two rooms in their own house to supplement their own limited incomes. Renting seems an essential part of the home consolidation process; without tenants, ownership would be more difficult for the landlords.
>
> (Gilbert, 1983: 472)

A similar pattern is evident in Caracas, Guatemala City, La Paz, Santa Cruz, and Mexico City (CEU, 1989; Rodas and Sugranyes, 1988; Beijaard, 1986; Green, 1988a; Gilbert and Ward, 1985). In Mexico City, Coulomb (1985b: 52) reports that 'the landlords of the urban periphery are ... labourers, employees, traders, bricklayers or artisans' and only one-third of the landlords had constructed the dwelling specifically to rent it.

Given our general ignorance of the nature of landlords and their operations, we attempt in this chapter to describe the owners of rental property in Guadalajara and Puebla. Our information is based on interviews with landlords in the two cities: ten in Guadalajara and thirty-seven in Puebla. For the reasons given in Chapter Two, this sample was obviously unrepresentative of all landlords in the two cities. Resident landlords with more than one tenant are over-represented in the sample while landlords with a single tenant are under-represented; in addition, too few owners of *vecindades* in Guadalajara were interviewed. The bias in our sample is evident from our other sources of information (such as the results of our tenant interviews and the household counts), which give a more balanced picture of the Guadalajara and Puebla landlord. Our discussion, therefore, is based partly on information from these other sources as well as on interviews with tenant and landlord organisations and rental administrators.

## LANDLORDS IN GUADALAJARA AND PUEBLA

### Size of operation

Our evidence strongly suggests that most landlords in the two cities operate on a small scale. Thirty of the landlords interviewed lived in the property where they let rooms; a further seven owned only one rental property. Half-a-dozen landlords owned two or three rental properties. Of course, larger operators exist: in Guadalajara, one landlord in Agustín Yáñez owned a *vecindad* for fifteen tenant households and at least five other properties; another, in Central Camionera, owned eight flats and a house for rent.[1] In Analco, Puebla, members of the Alonso family owned some seven rental properties, mostly *vecindades*, in addition to their own homes. Even in this case, however, individuals within the family owned no more than two or three properties.[2]

Further evidence from Puebla supports the idea that there are relatively few large-scale landlords in the city. Cadastral data on the central Analco area showed that few owners had more than one property. In addition, representatives of several rental companies agreed that most letting was on a small scale: the largest landlord on any of their books was probably receiving no more than one million pesos per month (at the time, the equivalent of about £1,000, or twenty times the minimum salary). This scale of holding was very unusual, and the most lucrative rental holdings were mostly in middle-class areas of the city. However, one caveat is necessary: the administrators admitted that the largest landlords in Puebla tend to manage their own property.[3] In Guadalajara, the overall picture of small-scale landlordism is generally similar. The municipal official in charge of the *vecindad* inspection programme reported that it was unusual for a landlord to own more than one property. Again, however, there must be a minor reservation: if small-scale ownership is indeed the norm, the situation would appear to have changed markedly since 1960. A survey conducted in that year recorded that 361 owners between them controlled 1,762 *vecindades*; 40 per cent of these properties were in the hands of forty-two owners, giving an average of seventeen properties apiece (Ramírez Jiménez, 1978).[4]

The apparent lack of large-scale landlords is linked to the absence of institutional landlords. In fact, we discovered only one example of the latter. This was a charitable medical foundation, created in the 1960s by a family with a considerable amount of rental housing in the centre of Puebla. The income from their *vecindades* was used to support a clinic run by the foundation. In the opinion of the current director, however, the rental properties are today more of a liability than an asset to the foundation. If this view is correct, it is not surprising that we found little evidence of businesses engaged in residential letting.

## Landlord residence and forms of rental housing

Some landlords live with their tenants; others live in the same settlement; a few do not even live in Mexico. In Guadalajara, few landlords lived with their tenants but in Puebla almost one-quarter of all tenants had a resident landlord (Tables 7.1, 7.2 and 7.3).[5] The resident landlord was found more frequently in the older-established self-help settlements of both cities than in the central areas. Too much should not be made of this distinction, however, for there were many more tenants with resident landlords in the centre of Puebla than in the older self-help settlement in Guadalajara. Among non-resident landlords, only a minority seemed to live in the same settlement; the majority of tenants said their landlord lived somewhere else in the city. Some landlords even lived outside the city, particularly in Guadalajara, where a number of landlords even lived in the United States. Such absenteeism was almost unknown in Puebla.

*Table 7.1*  Landlord's place of residence

| | GUADALAJARA | | | PUEBLA | | |
|---|---|---|---|---|---|---|
| | *Older settlement tenants* | *Central city tenants* | *All* | *Older settlement tenants* | *Central city tenants* | *All* |
| *Percentage of tenants interviewed reporting landlord living in:* | | | | | | |
| Same plot | 14 | 0 | 7 | 27 | 19 | 22 |
| Same area | 16 | 17 | 16 | 29 | 16 | 21 |
| Same city | 58 | 60 | 59 | 40 | 63 | 53 |
| Outside the city | 4 | 22 | 14 | 4 | 1 | 2 |
| USA | 8 | 1 | 5 | 0 | 2 | 2 |
| Total | 100 | 100 | 100 | 100 | 100 | 100 |
| Sample size | 71 | 77 | 148 | 73 | 97 | 170 |
| Percentage of tenants reporting that they did not know where landlord lived: | 26 | 36 | 32 | 9 | 16 | 13 |

*Source:*  Household survey.

*Table 7.2* Different types of rented property in the survey settlements

| | GUADALAJARA | | | PUEBLA | | |
|---|---|---|---|---|---|---|
| | Older settlement tenants | Central city tenants | All | Older settlement tenants | Central city tenants | All |
| *Percentage of rented plots with:* | | | | | | |
| Resident landlord + tenant(s)* | 6 | 0 | 3 | 26 | 21 | 23 |
| One tenant household only | 77 | 75 | 76 | 33 | 13 | 22 |
| Several tenants without landlord* | 17 | 25 | 22 | 41 | 66 | 55 |
| Total | 100 | 100 | 100 | 100 | 100 | 100 |
| Sample size | 142 | 194 | 336 | 173 | 205 | 378 |
| Rented plots as a percentage of all residential plots | 42 | 58 | 50 | 35 | 64 | 46 |

*Source:* A complete residential plot survey was carried out in the two rental settlements in Puebla. In Guadalajara, the plot/household listing was based on a sample of nine blocks (50 per cent of the total) in the older self-help settlement (Agustín Yáñez) and of eleven blocks (13 per cent of the total) in the central city settlement (Central Camionera). The small minority of plots for which insufficient information was obtained are excluded from the calculations.

*Note:* * Some plots also include households who are neither owners nor tenants – for example, sharers, or 'caretakers' in *vecindades*.

Landlords own a wide variety of types of property. Such variation is apparent both within the individual settlements and across the two cities; there are also significant differences between the two cities (Tables 7.2 and 7.3). Multi-occupancy buildings were the dominant form of rental property in Puebla. Only 4 per cent of tenants lived on their own, in properties accounting for 22 per cent of those visited. In contrast, in Guadalajara, two-fifths of all tenants lived on their own and three-quarters of all the property listed in our household counts was of this kind. Not surprisingly, large rental properties were more common in Puebla than in Guadalajara (Tables 7.3 and 7.4). In both cities, approximately one tenant in four lived in accommodation with ten or more other households (Table 7.3). There were more large properties in the central areas, although some were also

*Table 7.3*  Tenant households occupying different types of rented property in the survey settlements

| | GUADALAJARA | | | PUEBLA | | |
|---|---|---|---|---|---|---|
| | *Older settlement tenants* | *Central city tenants* | *All* | *Older settlement tenants* | *Central city tenants* | *All* |
| *Percentage of tenant households occupying rented plots with:* | | | | | | |
| Resident landlord: | | | | | | |
| – household listing | 15 | 0 | 6 | 34 | 20 | 24 |
| – qn. survey | 10 | 0 | 5 | 25 | 16 | 20 |
| One tenant household only: | | | | | | |
| – household listing | 39 | 29 | 33 | 10 | 2 | 4 |
| – qn. survey | 48 | 32 | 39 | 16 | 1 | 7 |
| Several tenants without landlord: | | | | | | |
| – household listing | 46 | 71 | 62 | 56 | 79 | 72 |
| – qn. survey | 42 | 68 | 57 | 59 | 83 | 73 |
| Total: | | | | | | |
| – household listing | 100 | 100 | 100 | 100 | 100 | 100 |
| – qn. survey | 100 | 100 | 100 | 100 | 100 | 100 |
| Sample size: | | | | | | |
| – household listing | 280 | 493 | 773 | 566 | 1368 | 1934 |
| – qn. survey | 96 | 120 | 216 | 80 | 115 | 195 |
| *Percentage of tenants living in properties with:* | | | | | | |
| 1 household | 48 | 32 | 39 | 16 | 1 | 7 |
| 2–5 households | 23 | 13 | 18 | 48 | 17 | 30 |
| 6–10 households | 19 | 19 | 19 | 18 | 32 | 26 |
| 11+ households | 10 | 36 | 25 | 15 | 37 | 28 |
| No. of house- holds unknown | – | – | – | 4 | 13 | 9 |
| All | 100 | 100 | 100 | 100 | 100 | 100 |
| Sample size | 96 | 120 | 216 | 80 | 115 | 195 |

*Source:* Household listing and household survey.

*Notes:* The percentage of tenants living on plots with a resident owner differs from that in Table 7.1 because of the different sample size. In some cases, it was known that the landlords did not live on the same plot, but their exact address was not known; these cases could be included in this table, but had to be counted as missing cases in Table 7.1.
Not all households living on a plot are bound to be tenants, except when there is only one household in all. 'No. of households unknown' almost always refers to cases where there are at least two households on the plot.
qn. = questionnaire.

found in the self-help areas. The largest properties that we found had fifty-nine tenants in Guadalajara and sixty-four in Puebla.

A further variation concerns the use of some rental properties to house members of the landlords' own family. In Puebla, various properties housed sharers as well as owners and tenants; this gives rise to a greater diversity of property types and tenure mixes than in Guadalajara.[6]

## Wealth

Some landlords are wealthy and some are rather poor; in our interviews we encountered a wide range of income groups. The causes of variation, however, are rather unclear and there seems to be no obvious relationship between the wealth of the landlord and variables such as location of the property, the manner in which people became landlords, or their place of residence.

However, the better-off landlords seem to have one thing in common: they tend to run their own business in addition to their rental interests. Indeed, it is this business that is usually the main source of their income. The family of one wealthy landlord in central Puebla owned a general store and a couple of businesses selling dress materials. They had built a block of sixteen flats at the front of their large property in Analco, and received approximately half their income from renting these flats. Another landlord in this area had a business exporting onyx craftware, which accounted for two-thirds of his income, the rest coming from ten flats which he had built himself. Most of these wealthier landlords owned property in the central area.

The majority of landlords, however, were not wealthy and some were decidedly poor. Perhaps the most extreme case was an elderly landlady who rented her own accommodation in a *vecindad* near Analco.[7] The most obvious link between these poorer landlords was that many were elderly, mainly older couples or widows. Many of the men were retired but others were still active, often in some kind of self-employment. The range of activities was wide, although commercial activities and skilled manual trades were frequently mentioned. There were also some younger widowed or abandoned women in this group of poorer landlords.

The percentage contribution of rent to these landlords' monthly incomes was highly variable. The more affluent landlords rarely reported that rent was the most important component, although the amount received from rent was none the less significant. Two landlords, with sixteen and thirty tenants respectively, received the equivalent of seven times the minimum salary, although the average for those with between ten and thirty tenants was around four times the minimum salary. A few of the more affluent landlords

*Table 7.4*   Size of rental properties in the two cities

| | GUADALAJARA | | | PUEBLA | | |
|---|---|---|---|---|---|---|
| | *Older settlement tenants* | *Central city tenants* | *All* | *Older settlement tenants* | *Central city tenants* | *All* |
| **Mean number of households living on plots with:** | | | | | | |
| Resident landlord + tenant(s)*: | | | | | | |
| – All | 4.8 | – | 4.8 | 4.2 | 6.2 | 5.2 |
| – Tenants only | 3.8 | – | 3.8 | 2.8 | 5.0 | 3.9 |
| One tenant household only: | | | | | | |
| – Tenants only | 1.0 | 1.0 | 1.0 | 1.0 | 1.0 | 1.0 |
| Several tenants without landlord*: | | | | | | |
| – All | 5.3 | 7.1 | 6.5 | 4.5 | 7.9 | 6.7 |
| – Tenants only | 5.3 | 7.1 | 6.5 | 4.5 | 7.8 | 6.6 |
| *All rented plots:* | | | | | | |
| – All | 2.0 | 2.5 | 2.3 | 3.3 | 6.7 | 5.1 |
| – Tenants only | 1.9 | 2.5 | 2.3 | 2.9 | 6.3 | 4.8 |
| No. of plots | 142 | 194 | 336 | 173 | 205 | 378 |

*Source:* Household listing – see Table 7.2.

*Note:*   * The number of households includes the owner, or households with other types of tenure, where present.

claimed to receive less than 10 per cent of their income from rent, but the figure was more often between one-half and one-quarter. As might be expected, more of the poorer landlords were completely dependent on the rent. However, the total income derived from rent was seldom substantial and it did not generally constitute a high proportion of landlords' monthly income. In Guadalajara, landlords with between two and six tenants received on average the equivalent of about one-and-a-half times the minimum salary from rent; in Puebla, the average for poorer landlords, with similar numbers of tenants, was less than one minimum salary.[8]

**The origins of renting**

There are various routes into the business of renting: some landlords bought property, some inherited it, and others built it themselves. While these routes into landlordism would seem to be very different, they often overlap; indeed, a few landlords had acquired property through a variety of methods.

A few landlords had bought property with the express intent of letting it. Among these were a number who were getting on in years and who wanted a secure income to support them and their families in their old age. One couple had bought a *vecindad* property in Agustín Yáñez when the man was 63 years old. He had made some money from agriculture and intended the income from the property to support his wife and himself through their old age. Lacking children or other relatives, this was needed as the equivalent of a pension. In hindsight, he thought that he had made a mistake; he could have gained a more secure income through putting his money in the bank.

The second and seemingly most common route into landlordism is through inheritance: most of the landlords with whom we talked had had the property, and the associated problems, wished upon them. Indeed, in central Puebla a number of properties had been passed down through several generations. One elderly landlady had inherited a *vecindad* from her grandfather who had built the property after he had migrated from rural Oaxaca in the mid-nineteenth century. Another landlord, who operated the worst *vecindad* we encountered, had also inherited it from his father; the father had in turn inherited it from his parents.

Renting in Mexico is often described as a 'widow's business'. The term derives from the time when property was a good investment and when a widow could live well from inherited property. Today, however, the term is also used in a more derogatory sense: it is only widows that remain in the business. Whatever the precise meaning, a few landlords did buy property with the intention of leaving it to their wives when they died.[9] Others bought it for their children, as something to pass on. One man of 69, for example, had bought a fifteen-family *vecindad* in the centre of Puebla for his daughter.[10]

Finally, there are the landlords who have built property themselves. A landlady in Puebla now had thirty tenants in three different properties in the city. Over the years she had bought land and gradually constructed flats to let. Other landlords had inherited land or property and later extended the business. A middle-class landlord in the centre of Guadalajara had partially converted the house he inherited from his mother and rented out four flats. He had also built flats elsewhere in the city. We found few other landlords, however, who had invested in new purpose-built accommodation.

The only substantial constructors were what we have called the 'self-help landlords'. These 'self-help landlords' normally began by building their own home and then gradually extended the property. Such extensions often took a long time and we interviewed several landlords who had purchased property in the 1950s, but not begun to let until the 1970s or 1980s. Clearly, for most of them, extending the property for rental purposes was a method of supplementing the family income. However, the strategy of building accommodation for rent in order to finance improvements to the family home was not common: we encountered only two landlords who had begun to rent within a couple of years of purchasing their plots. Indeed, we suspect that many of these self-help landlords drifted into landlordism. Certainly, several of them had only begun to let rooms once their children had grown up and left home. A few are undoubtedly drifting out again: we found several cases where the tenants had been evicted in order to make room for grown-up children and their families. We also found several plots with a combination of landlords, sharers and tenants.

Among these self-help landlords were several who had inherited plots from their parents. This was especially common in the central area of Puebla, where one part of Analco was not developed until the 1940s, but it was not uncommon even in the consolidated self-help settlement.[11] This underlines the point that the different routes into renting frequently coalesce. Self-help landlords sometimes inherit property; middle-class owners extend property that they have inherited; some landlords purchase a plot and build upon it. Whatever their route into landlordism, however, few believe it to be a profitable investment.

## THE ECONOMICS OF LANDLORDISM

### Rent levels

In Mexico, conflicting claims are constantly being made about rent levels; tenants claim rents are too high; landlords, that they are too low. Without getting too far into the question of what constitutes the basis for a fair and equitable rent, it is necessary to establish certain facts about rent levels in the two cities in 1985 and 1986.

First, to judge by international standards rent levels in Guadalajara and Puebla were not high relative to household incomes. The survey data show that the mean proportion of income going in rent was 13 per cent. Even for those earning less than one minimum salary, the mean rent/income share was only 16 per cent. By comparison, World Bank estimates for thirteen cities in less-developed countries demonstrate that rent/income shares vary from a high of 22 per cent in Seoul to a low of 7 per cent in Cairo. Table 7.5 and Figure 7.1 show that if the rent/income shares of those earning less than

US $ 50 is considered, the range is from 77 per cent in Seoul to 9 per cent in Davao (Malpezzi and Mayo, 1987b). Unlike income/expenditure patterns, average rent/income shares rise as cities become more prosperous: 'as income increases from Cairo, to Manila, to Bogotá, and then to Seoul – so, too, does the average fraction of income allocated to housing' (Mayo, 1985: 5). If this interpretation of the data is correct, the housing/income shares of Guadalajara and Puebla should fall between those of Bogotá and Seoul. In fact, the mean rent/income ratios were considerably lower than those of Bogotá.

To take comparison further, rent/income data from Guadalajara and Puebla have been compared in Table 7.6 to those found in similar kinds of settlements in Bogotá (Gilbert, 1983). In the latter, rent/income shares seem to be somewhat higher. If we look at the figures by income group, the shares are much higher for the lowest three income groups in Bogotá. This is especially interesting in so far as the Guadalajara and Puebla figures were collected during a recession when the rent/income shares might be expected to have risen; the comparison suggests that rents in the two cities were relatively low.

It is also interesting to compare the latter figures with the rent/income

*Table 7.5*   Rent/income shares by city and income group

| | Percentage of households earning US dollars per month: | | | | |
|---|---|---|---|---|---|
| | 50 | 100 | 150 | 300 | City average |
| Seoul (Korea) | 77 | 52 | 42 | 29 | 22 |
| Kwangju (Korea) | 46 | 35 | 30 | 23 | 21 |
| Cali (Colombia) | 47 | 32 | 25 | 17 | 19 |
| Taegu (Korea) | 53 | 36 | 28 | 19 | 18 |
| Bogotá (Colombia) | 33 | 26 | 23 | 18 | 18 |
| Busan (Korea) | 68 | 42 | 32 | 20 | 16 |
| Bangalore (India) | 12 | 9 | 8 | 6 | 10 |
| Beni Suef (Egypt) | 11 | 8 | 6 | 5 | 9 |
| Manila (Philippines) | 23 | 17 | 14 | 10 | 9 |
| Santa Ana (El Salvador) | 17 | 12 | 9 | 7 | 8 |
| Sonsonate (El Salvador) | 15 | 11 | 9 | 6 | 8 |
| Davao (Philippines) | 9 | 9 | 8 | 8 | 8 |
| Cairo (Egypt) | 10 | 7 | 6 | 4 | 7 |

*Source:*  Malpezzi and Mayo (1987b: 210).

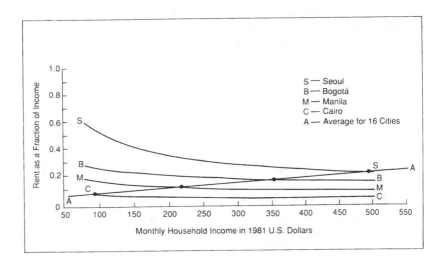

*Figure 7.1* Rent/income shares by household income in ldc cities

*Source:* Mayo, 1985.

shares in Mexico City in 1970. After a decade when rents had risen 50 per cent faster than prices in general, almost half of middle-income families were paying between 0 and 10 per cent of their income in rent and a further 39 per cent between 11 and 20 per cent (Slovik, 1972).

Second, the general proposition that rent levels were not high in Guadalajara and Puebla in 1985 and 1986 is also supported by the relationship between changes in rents and prices between 1970 and 1988. During that period, in fact, rent levels rose more slowly than prices in general. Table 7.7 shows that the costs of housing (rents, fuel and light) broadly rose in line with prices, except in years of rapid inflation. Whenever the general rate of inflation was rising rapidly, the ratio between housing costs and prices appears to have fallen. This was most obvious in the periods 1973–4, 1976–7, and 1982–7. The most dramatic change in the ratio occurred during the rapid inflation of the 1980s; in 1981, the two indices were more or less equal; by July 1987 the general index was almost twice that of the housing index.

It would be misleading, however, to compare housing costs only with the general price index. Housing costs should also be compared to levels of income, for despite the overall trend it is possible that rents have risen more rapidly than incomes. Table 7.7, therefore, attempts to compare trends in

*Table 7.6*    Rent/income shares by income group in Guadalajara, Puebla and
Bogotá (per cent)

| Income groups | GUADALAJARA | | PUEBLA | | BOGOTA |
| | Older settlement tenants | Central city tenants | Older settlement tenants | Central city tenants | Self-help settlements* |
| --- | --- | --- | --- | --- | --- |
| 1 | 26 | 22 | 27 | 10 | 42 |
| 2 | 13 | 19 | 12 | 12 | 20 |
| 3 | 9 | 10 | 9 | 9 | 14 |
| 4 | 7 | 6 | 8 | 4 | 9 |
| 5 or over | 7 | 5 | 3 | 3 | 7 |
| Sample size | 70 | 69 | 52 | 67 | 105 |

*Source:*  Household survey and Gilbert, 1983.

*Notes:*    Household income ranges are: 1 = 0–1.0 minimum salary, 2 = 1.01–2.0 minimum
salaries etc. Rent/income percentage is the mean value for observations in the
corresponding income group.
* Four self-help settlements which cannot be disaggregated as a result of small
sample size.

rents with the minimum salary. The data show that although there was a
general decline in the housing/minimum salary ratio during the 1970s,
during most of the 1980s the ratio was fairly steady.

It is only after the *Pacto de concertación* was introduced in 1988 that
rent levels began to rise substantially faster than the minimum salary. As a
result of the government's anti-inflation policy, the minimum salary was
permitted to rise by only 11 per cent during 1988. In contrast, and despite
the aim of the government, rents rose by 84 per cent. Despite the limits to
rent rises laid down in the legislation operating in several Mexican cities,
landlords continued to raise rents by roughly the same amount as they had
during the previous six years of rapid inflation.

A glance at Table 7.7 suggests that recent rises may have been
exceptional. Past experience has been that landlords have failed to raise
rents in line with changing prices when the general inflation rate has
suddenly accelerated. Equally, they have failed to limit rent rises when the
rate of price rises generally has fallen. In fact, there seem to be signs of a
lagged response by landlords to the general rate of inflation. If this is
correct, rent rises will soon move back in line with general rises in prices.

There are several ways of explaining the trend up to 1988. First, if the
Mexican rental legislation is effective – and we are by no means sure how
effective it is – then rents are bound to fall relative to incomes during a
period of rapid inflation. During an existing contract landlords are not

*Table 7.7*   Inflation, housing costs and minimum salaries, 1970–89

| Year | Rate of inflation (per cent) | Housing cost/ general price index (July) | Housing cost/ minimum salary (July) |
|------|------|------|------|
| 1970 | 3.8 | 1.16 | 1.39 |
| 1971 | 4.2 | 1.16 | 1.46 |
| 1972 | 5.0 | 1.17 | 1.30 |
| 1973 | 12.0 | 1.13 | 1.41 |
| 1974 | 23.8 | 1.02 | 1.16 |
| 1975 | 15.2 | 1.01 | 1.08 |
| 1976 | 15.8 | 1.03 | 1.04 |
| 1977 | 20.7 | 0.97 | 0.93 |
| 1978 | 16.2 | 1.00 | 1.00 |
| 1979 | 20.0 | 0.99 | 1.01 |
| 1980 | 29.8 | 0.96 | 1.05 |
| 1981 | 28.7 | 0.97 | 1.05 |
| 1982 | 98.9 | 0.93 | 1.19 |
| 1983 | 80.8 | 0.82 | 1.13 |
| 1984 | 59.2 | 0.75 | 1.10 |
| 1985 | 63.7 | 0.70 | 1.04 |
| 1986 | 105.7 | 0.68 | 1.13 |
| 1987 | 159.2 | 0.58 | 1.03 |
| 1988 | 51.7 | 0.60 | 1.27 |
| 1989 | 19.7 | 0.77 | 1.70 |

*Sources:* Banco de México *Indice Nacional de Precios al Consumidor*: cuadro III–1.
Comisión Nacional de los Salarios Mínimos: *Salarios Mínimos*.

*Notes:*   Inflation      – annual figures for 1 January – 31 December.
Indices        – 1978 = 100.
Minimum salary – average minimum salary for country on 1 July each year.
Clearly, rises in the minimum salary are somewhat jerky
so that there are sudden variations in the ratio when the
minimum salary changes.

permitted to raise the rent by more than a set proportion of the rise in the minimum salary. Since the minimum salary has been increasing more slowly than prices, this would be sufficient to explain the fall in rents relative to prices in general. It would not, of course, explain why rents rose marginally faster than the minimum salary during most of the 1980s and certainly not the major jump in 1988 and 1989.

A second explanation of the slow rises before 1988 is that many landlords put up the rent of current tenants with some reluctance. While many raise the rents when the annual contract is renewed, others fail to do so. Many landlords do not like discussing rents with tenants especially when they have known them for some time; some shy away from dealings with recalcitrant tenants. The only time when landlords are free of such constraints is when a tenancy is terminated. This distinction would be of little significance were it not for the fact that the median current tenancy in our survey settlements is lengthy – it varies from a minimum of three years in the older self-help settlements to eight or nine years in the central areas (see Chapter Eight). If landlords fail to raise the rents of existing tenants in line with prices generally and those tenants stay in the property for several years, the real value of rents will fall. During a period of rapid inflation, it does not take long for rents to drop behind the general price level and behind the general income level. Such an argument is supported by cross-national findings: Malpezzi and Mayo (1987a: 705) report that 'length of tenure and housing expenditures are negatively related in all estimates'. It is also compatible with practice in Los Angeles where rent rises for new tenants are far higher than those of existing tenants (Malpezzi and Rydell, 1986: 11).

Third, even when the contract is terminated and a free market rate can be negotiated, it is less than certain that landlords obtain rises equivalent to those of prices in general. As one rental administrator in Puebla argued, rents cannot be raised continuously because people simply do not have the money. This would certainly explain the implicit link shown in Table 7.7 between rises in rents and the minimum salary. It is also compatible with the impressions gained from interviews that, while some landlords extract all they can from new tenants, many landlords are much less economically rational. One Puebla landlord claims to set the rent according to the tenants' means; another puts it up a little to cover small repairs to the flats; others 'occasionally' put up rents when tenants change. Again, however, this explanation does not account for the changes after 1987.

Finally, lowish rents are compatible with the trend towards owner-occupation in most Mexican cities since 1940. Elsewhere we have shown that the cost of unserviced peripheral land relative to incomes is quite low in Guadalajara and Puebla. Earlier information for Mexico City also suggests that *ejido* land was relatively cheap compared to plots in pirate subdivisions in Bogotá (Gilbert and Ward, 1985). If this is the case, then it is logical to believe that rents will also be low. If rents were raised in line with inflation, larger numbers of tenants would opt to become owner-occupiers on the periphery. Of course, such a change cannot be made instantly, but there ought to be a general relationship between the cost of peripheral land and the level of rents.

Of course, all this information is circumstantial. It also conflicts with some of our other findings, notably that there is no shortage of tenants in Guadalajara and Puebla. Our argument also diverges from the claims of several studies that rents in Mexico City are increasing more rapidly than inflation (Portillo, 1984; Méndez Rodríguez, 1987), although since 1988 rents clearly have been doing so.[12] None the less, we believe that the evidence, such as it is, supports the idea that rents up to 1988 were low both relative to prices and relative to incomes. This is not, of course, the same as saying that poor households find it easy to pay the rent. Under a generally inflationary situation and with falling real incomes, they are faced with major budgetary problems. It is, however, compatible with what the majority of landlords were saying during 1985-6: that renting was not a good business.

## The economic returns from letting

Few of the landlords interviewed in Guadalajara and Puebla are likely to become wealthy on the strength of their rental income. The maximum received by any of the landlords was seven times the minimum salary.[13] The majority of landlords interviewed received the equivalent of less than one minimum salary from rent.

Out of these rents, landlords have certain expenses to cover, principally the cost of water and property taxes. In fact, such outgoings are generally modest and most landlords had considerable difficulty in remembering how much they had to pay. Only a couple of landlords mentioned them as major problems. Landlords living on their property would in any case have to pay the same amount in property taxes whether or not they had tenants. Some landlords should also be liable to income tax on their rental income, although few, if any, of the landlords actually pay it.[14] The only major expense, therefore, is the cost of maintenance, although many landlords get round this problem by failing to repair their properties.

With receipts often amounting to little more than a minimum salary from renting, it is very difficult to see how building a new rental property in low-income areas of the city can be a sound business proposition. It is almost equally difficult to see how buying an existing property can give a reasonable rate of return on investment. The opinions expressed by landlords amply confirmed this view. The great majority were absolutely clear: renting is bad business. Typical comments were: 'no, you couldn't really call it a business', 'no, it's not a business', 'no, no, no! Right now, renting houses is the worst, really the *worst*, business you could name', and 'no, it's the worst business I've ever got involved in'. One landlord, who was very unusual in having bought his rented property within the last year, admitted that 'I've put my foot in it'. Renting was considered a poor business mainly because rents are low, but landlords also complained that maintenance costs were so high that they couldn't afford to repair their property. The lack of repairs means that the property deteriorates over time, ensuring that rents *remain* low. Renting is not only bad economics: it is also an unpleasant and inconvenient business. Collecting the rent is a headache both because it is a physical chore and because of the difficulty of extracting the money from tenants who do not pay on time. Putting up with aggressive tenants, and getting rid of bad tenants, are difficulties which landlords could do without. As one landlord said: 'I'd get three times as much if I had the money in the bank, and the cheque would be delivered to me here, no problems'.

Within this general chorus of disgruntlement, there are a few dissenting voices. One landlady admitted that low rents made letting a poor business, but noted that it was a secure source of income, and 'any help is a good thing'. This sense of the income from rent 'helping' in some, albeit minor, way was conveyed by all eight of the landlords who dissented from the generally very poor opinion of renting. The main advantage was the guarantee of receiving some money in the form of rent. Tenants would always require accommodation, and so renting 'is a business which doesn't end'. It was especially valued by old people without any other regular source of income. Help from relatives cannot be counted on by *all* old people, and even some of the reluctant landlords acknowledged that renting had certain attractions in this respect. The prospect of rents providing an income, however low, for one's old age is one reason why many landlords continue to rent out property. The traditional saying that 'I'm going to retire, and live off my rents', is still partially relevant.

Other landlords explicitly acknowledged their lack of business logic in continuing to rent. Some admitted that they kept their property mostly for sentimental reasons. One landlord in the centre of Guadalajara keeps his inherited property 'in memory of my parents'.[15] Another landlord kept his colonial *vecindad* in Analco mostly as an inheritance from his parents, and as somewhere to live for himself and his family. He actively disliked having strangers on the plot and intended to reduce the number of rented rooms if and when economic circumstances allowed. Others kept the property so that they would have something to leave to their children.

Landlords were also commonly following a now outmoded business rationale – the 'traditional' belief, among those with money to invest in Mexico, that it should be put into real estate rather than into banks.[16] The persistence of such a belief in the face of evidence to the contrary is more readily understandable when it is remembered that most landlords are drawn from an older generation. The few landlords interviewed who had invested in property in recent years were old men. They were also from relatively poor backgrounds, working, for example, in agriculture. They were clearly not shrewd businessmen.

Approximately one-quarter of the landlords interviewed intended either to reduce the number of rooms they were renting or to sell the property; a number of them were in fact already trying to do so. In practice, the difficulties they were encountering in selling suggest that one reason why landlords continue to rent is that they are trapped into it. One man had tried unsuccessfully to sell his *vecindad* in Agustín Yáñez; a woman who had inherited property in the same settlement was also trying to sell. One landlord owning property in Central Camionera had already sold one of his four flats to an employee; he wanted to sell all the property and thought it good for his employees to have the responsibility of home ownership.[17] Landlords probably have a better chance of selling to tenants than to other landlords. Several families in a *vecindad* in Analco had successfully come to an agreement with their landlord to buy their homes; at least one other landlord in the same area was reported to be selling off flats to the tenants. However, there is a major problem: few tenants have the money to buy outright and bank loans are both very expensive and very difficult to obtain.[18] Landlords are also unwilling to let tenants pay in instalments, because of the high level of inflation. As a result, although both sides may be willing to transfer ownership, the situation often becomes a stalemate.

Some landlords are less interested in selling the property to their tenants than in refurbishing it and selling it off as a condominium. To do this, however, they first have to evict the existing tenants. This sometimes proves difficult. In central Puebla, a young landlord was planning either to sell his property of six flats or pull it down. Three of the six tenants had already left, but he was involved in a conflict with the others, who neither wanted to leave nor to buy their flat from him. He was worried that they would take him to court, which would mean he had to wait even longer before he could get them out and convert his property for sale.

In seeking to convert their property into condominiums, landlords were following a general trend of rental disinvestment. Rental administrators and representatives of landlord organisations interviewed in the two cities admitted that selling property was more profitable than renting it. Although the organisations concerned administered property in mostly middle-class areas, they reportedly found renting just as bad a business as the landlords of property in the case-study areas. The return from renting, which traditionally had yielded around 1 per cent of the property's value per month, was now producing only 0.2 per cent. Owners generally were 'pushing to sell', mostly by converting and selling off rented flats for owner-occupation or for offices. Although some rental property was still coming on to the market, and although the agencies were still getting some new owners as clients, the administrators were agreed that the current climate was one of strong disinvestment.[19]

The conversion of property into condominiums is profitable providing that the property is sufficiently attractive to attract purchasers, that credit is available for buyers without ready cash, and that the owner can mobilise the capital to cover the costs of conversion and improvements. Currently, however, credit is very expensive and often unavailable. This has left the owners of ageing rental property, particularly *vecindades* in the city centre, with no real option but to try to sell the property as it stands. A number of *vecindades* in each city carried signs advertising their sale, and occasionally properties were advertised in the local newspapers. In general, however, rental property was not moving. A survey of ownership changes in Analco revealed a generally low level of turnover of properties; the results are summarised in Table 7.8. Almost three-fifths of the plots in Analco have not changed hands since 1970; and approximately one-fifth of them have not changed hands at all since 1950.[20] The conclusion is that ownership patterns are very stable.

If rental property cannot be sold, then the only other alternative is to sell it for its site value. This is an attractive option for landlords with property in the central area; it is certainly a better option than renting. One building in central Puebla, with twelve homes and four shops, which was reported to yield 70,000 pesos (less than twice the minimum salary) was on sale for 42 million pesos; another, with a similar number of 'flats' but only two shops, was advertised at 36 million pesos. Even in the central area, however, such sales could not be guaranteed, especially for plots of less than 500 square metres. For smaller properties, prices were much lower: one block of four flats yielding 85,000 pesos per month was advertised for sale at only 7 million pesos.

Selling vacant property, however, still requires eviction of the tenants, which is not an easy task. Another method for resolving these dilemmas has therefore emerged: the landlord simply allows the property to fall down around the tenants' ears. This is an extreme solution, but one which was being used in both Central Camionera and Analco. In both areas, *vecindades* were found in an advanced state of decay. The roof to individual rooms had often fallen in; vacant rooms were filled with rubbish and rubble, and were frequented by rats. The owners did not necessarily trouble to hide what was happening to their properties, since they would even allow the rooms opening onto the street, often used as small shops, to fall into the same state of disrepair. The problem is most severe in Puebla, where the historic monument legislation which has supposedly protected colonial and nineteenth-century properties since 1977 may actually have had the opposite effect. The legislation severely restricts permissible changes of

*Table 7.8* Changes of plot ownership in Analco, 1950–86

| Number of changes of ownership known to have occurred since: | Percentage of plots: | | | |
|---|---|---|---|---|
| | *1950* | *1960* | *1970* | *1980* |
| 0 | 21 | 34 | 57 | 87 |
| 1 | 42 | 42 | 34 | 12 |
| 2 | 24 | 16 | 7 | 2 |
| 3 | 10 | 4 | 2 | 0 |
| 4 | 2 | 1 | ng | 0 |
| 5 or more | 1 | 2 | 0 | 0 |
| Total | 100 | 100 | 100 | 100 |
| Number of plots | 220 | 305 | 302 | 312 |

*Source:* Data from the *Dirección de Catastro, Secretaría de Finanzas del Estado de Puebla.*

*Note:* Care should be exercised when comparing data across the years because the number of missing cases varies, particularly for the 1950s: thus, a property whose earliest recorded change of ownership is, for example, 1956, is lost from the 1950 column, and if the owner in 1978 is not the same as the owner recorded in 1964, without adequate information being available to date the change between these years, the plot will be included in the base for the 1960 and 1980 columns, but not the 1970 one.

use, modifications to the fabric of the building etc., thereby hindering the type of conversions in which owners wish to invest.[21] Many colonial *vecindades* in Analco and other parts of central Puebla are therefore standing empty, in a state of semi-dereliction.[22] One study of an area in the centre of Puebla which contained 645 'listed' properties (489 of which were residential) revealed that forty-five were wholly or partially abandoned, and thirty-four of these were already in ruins or in imminent danger of collapse.[23] In Guadalajara, fewer semi-derelict buildings are to be seen from the streets. Nevertheless, a register in the Municipal Planning Department shows that 109 out of a total of 1,137 *vecindades* were abandoned, and a further thirty-five almost deserted. Visiting some of the addresses in Central Camionera which had been registered as *vecindades* in the 1970s revealed some of the changes of use which had taken place since then: one plot was now used as a school; another for warehouses; a third was used partly as offices and partly to repair cars. In the end, most of these properties will either be demolished or be converted to other uses, although one colonial *vecindad* in Analco was of such architectural interest that a government body was renovating it.

The conclusion to this discussion would appear to be that rental housing in the central areas, and particularly the *vecindad*, is a form of accommodation which, unless there is a radical change in economic circumstances, is doomed to further decline. Rents are too low to allow adequate maintenance, and many properties would be more attractive if they were cleared and sold as land – or even used as parking lots. In short, the central rental sector appears to be a residual one, with no prospects for success, at least as far as low-income accommodation is concerned. The poor prospects for this sector were well summed up by one man who had been involved in the declaration of central Puebla as a zone of historic and architectural interest: 'the historic centres', he said, 'cannot survive as low-income housing zones'. In his opinion, the best solution would be to 'repopulate the area with another type of people'.[24] This chilling vision of the future is one which does not seem, to date, to have evoked any significant political response on the part of the inhabitants of the central areas.

The owners of *vecindades* in the older self-help settlements would seem to be equally disillusioned with renting and would get out of it if they could. The same does not apply to 'self-help' landlords, all of whom intended to continue renting, at least for the time being. While many complained that it was a bad business and declared that they would not invest money in building rooms to rent in the future, they did not intend to dispose of their property. Some landlords will probably continue to build housing to let. For most of this group, putting money into bricks and mortar is the only type of investment they know anything about. Most of them were once tenants and would share the sentiments of one landlady who declared 'I've come up from nothing'. Just as they built their own house little by little, they can add on extra rooms for rent. What is more logical than to put a little spare cash into extra housing space? Their expectations in renting are more modest than those of a commercial investor; they are therefore less likely to be disappointed by the results. And, even if they do have trouble with some tenants who do not pay the rent, or indulge in rowdy behaviour, what else are they to do with their extra rooms until their children are old enough to need them? They might as well look for a new tenant and hope for better luck next time.

The majority of the 'self-help landlords' sooner or later mentioned their children. A number planned eventually to hand the rented rooms over to them. In other cases, if their children got married and needed somewhere to live for a while, they were prepared to put them up. If they depended very heavily on the rent for their income, they could always charge their children rent. For some, accommodating their offspring was a relief from their dissatisfaction with renting, as in the case of one landlady who had experienced problems with tenants not paying the rent and was gradually installing her children in rooms previously occupied by tenants. In others, it was a response to unexpected circumstances: one landlady in Veinte de Noviembre had added three independent flats to her home, when some of her children had found themselves in financial difficulties. In this case, it was likely to be a temporary situation 'how good it would be if they all went to live independently; we could have more tenants if they did!' It can be seen from this that the replacement of tenants by married children is not a one-way process.

In practice, many landlords were replacing tenants with their own relatives. Eight per cent of tenants in Agustín Yáñez, and 18 per cent of those in Veinte de Noviembre, said they were relatives of the owner. In addition, there were large numbers of grown-up children sharing, rent free. This process could lead, either to a situation in which owners, tenants and sharers all lived on the same plot or to one in which the landlord ceased to rent altogether, and the plot was inhabited only by owners and sharers. Together, these two types of plot could actually outnumber the plots occupied by owners and tenants: in Veinte de Noviembre, they accounted for 12 per cent of all residential plots, compared with only 7 per cent of plots with a resident owner and tenants.[25]

It is this flexibility of use that perhaps explains why so many of the 'self-help landlords' continue to let accommodation. They are even continuing to build: new landlords were emerging in El Salvador, although the settlement was only a few years old. At the same time as rental properties in the centre of Puebla were being allowed to fall down around their tenants' ears, houses in El Salvador were being rented out. Even though a large proportion of the settlement area was still uninhabited, 5 per cent of the residents were already tenants. It is this development of renting hand-in-hand with owner-occupation that provides the only real source of new accommodation for the poor in Mexican cities.

## CONCLUSION

Our intention in this chapter has been to provide some kind of portrait of the landlord. This has been necessary because there has been so little work on Mexican landlords. With the exception of scholars such as Coulomb (1985b) and Marroquín (1985), most descriptions of landlords in Mexico contain more myth than reality. The picture that emerges from our study is not a totally clear one, for it is obvious that landlords in Guadalajara and Puebla are highly diverse, both in their origins and in their form of involvement in renting. The majority of landlords are clearly operating on a small scale; they own only one rental property and may not be conspicuously wealthier than their tenants. Some landlords, however, are more affluent people with a greater number of properties and tenants (although their relative affluence is more likely to be derived from business interests or employment than it is from renting). What links these different types of landlord is that few find letting a highly profitable activity; indeed, many would like to get out of the business. While there is a group of 'self-help' landlords who are continuing to invest in rental accommodation, many others have become landlords by chance: they are landlords because they have inherited rental property. Some landlords of inherited property are unwilling to dispose of it for sentimental reasons, and, like many 'self-help' landlords, they may value the opportunity their property gives them to help their children make a start in married life. Economics is not, therefore, the only consideration taken into account by landlords in deciding whether to continue to let property.

## APPENDIX: LANDLORD CASE STUDIES

The following case studies are chosen to provide illustrations of the nature of landlords in the two cities and the kinds of problems they face.

### 1   Analco, Puebla

Delia Patiño is a wealthy landlady who lives on more than 5,000 square metres of land, occupied partly by her own house and large garden, and partly by eighteen rented apartments. It is her only property. The whole block was acquired by her mother in the 1940s, but part was sold to a soft-drinks bottling business which still occupies the rest of the block. There were already nine small flats when the property was purchased; Delia's mother had the rest built around 1950. The main house is fairly well-maintained, but the rental property is ageing and not in very good condition. Delia inherited the property from her mother in 1985. She is a divorced woman in her forties; two of her four children live with her, and are still at school. Delia works in a travel agency belonging to a friend, earning rather more from her job than she receives each month from the rents. Disregarding anything she may receive from her ex-husband, her income comes to about eight times the minimum salary. At one point in the past, she was herself a tenant, and this makes her rather intolerant with her tenants when they fail to pay the rent on time. When she was a tenant, she said, the first thing she used to do each month was to put aside the money to pay the rent. In general, she has 'no patience' with renting, and she is trying to sell off the property to her existing tenants. She has had no success to date.

## 2 Central Camionera, Guadalajara

Gabriel Zarate owns a corner plot on which there are four houses. He inherited the property from his father, some eleven years ago, and keeps it mostly for sentimental reasons: it is 'in memory of my parents'. He lets two houses, each occupied by a single family, and allows the family of one of his three children to occupy a third. Another married son lives above Gabriel's own house, which he now occupies with his wife. Gabriel is middle-aged and owns a small business, making machinery for shoemakers. The business provides the bulk of his income. The rents amount to just over one minimum salary, which he doesn't 'take ... into account'. Since he is reasonably well-off he is not troubled by the low income he receives from renting: he is not renting, he says, for economic reasons.

## 3 Central Camionera, Guadalajara

Margarita Contreras is a widow in her late sixties. She has eight children, none of whom now live with her; instead, a couple of other young relatives share her home. She is comfortably well-off, if not ostentatiously wealthy. She lives in a neat property next door to an equally well-maintained block of four flats which she rents out. She also has eleven flats in a middle-class suburb in the west of Guadalajara; one of these is occupied, rent-free, by a daughter. Her family also own six small commercial and industrial properties, including a small electrical goods and repair shop on one of the city's main shopping streets. The rented houses were built by Margarita's husband, one at a time, from the late 1950s onwards. He died seventeen years ago, leaving this property to his widow. The rents from the flats amount to about twice the minimum salary. She lives entirely on these rents and on those from the business properties. Her children help her collect the rents.

## 4 Analco, Puebla

Tomás Benítez owns a split-level property on the main street in Analco. At the front are two shops. He lives upstairs around an open patio shared with the three tenant households. Tomás is clearly not wealthy. He comes from a small town nearby, and travelled widely throughout Mexico before coming to live with his parents in Puebla. He inherited this property when his first wife died, and now lives there with his second wife and their seven children, aged from 1 to 20. He is a mechanic, working for the Volkswagen garage in Cholula, and earns the minimum wage. The income from the three flats comes to just over half a minimum salary and he receives more from the two shops downstairs. Although renting is not very remunerative, 'any little helps'. The rooms are all old, and they are in need of repair; they are unlikely to be done up in the near future.

## 5  Analco, Puebla

Josefina Aguilera is very poor; indeed, she is herself a tenant, living in two rooms in an old *vecindad* near Analco. She is a widow, 86 years old, very slight and frail, and hardly able to get about outside the house. She owns a small house with four flats in Analco, which she inherited from her husband when he died in 1980. They had been married for 55 years but had not had any children; she now lives alone. Josefina has a brother in Mexico City, but does not want to live with him. For a time, she lived with a godson in a village near Puebla, but did not get on well with his family; she therefore moved to her present home. She pays a woman friend to look after her at night, in case she is taken ill; this friend sometimes collects the rent for her.

Her husband bought the rented property in 1970, with the intention of providing for himself and his wife in their old age. The couple had been shopkeepers in Los Remedios, near where she now lives; but Josefina sold the shop. Her rooms in the *vecindad* are reasonably well-furnished, but she says that she can only pay her own rent if the tenants pay on time. If they get behind with their payments, she sometimes has to borrow money. She finds it very difficult to get them to accept higher rents, although she believes that she could get twice as much money for the flats as she is doing at present. The total income from the rent is the equivalent of three-quarters of a minimum wage.

## 6  Veinte de Noviembre, Puebla

Pablo Núñez and his wife Amelia own one property, where they live with two tenant households and two of their married children. They are an elderly couple, with six other grown-up children. Pablo is a retired textile factory worker, receiving a pension worth about half the minimum salary. Both he and his wife still work as traders, selling a variety of goods. They have not been doing very well recently, as Amelia quickly gets tired. As a result, they depend fairly heavily on the rents from the two rented flats, which total about three-quarters of the minimum salary. Occasionally, they have been unable to buy food because their tenants have not paid the rent.

Before moving here, the couple had always lived in *vecindades*, which they hated; they jumped at the chance to buy this plot, 35 years ago. They built one room plus a kitchen and came to live in Veinte de Noviembre straight away. Over the years, they gradually added the rest of the building, including the two small flats now occupied by the tenants. The second rented flat had taken five or six years to build, because they were short of money. Eventually, under pressure from people wanting to rent the unfinished rooms, they had taken out a loan to complete the work.

Renting is not a good business. Pablo wanted to sell up but Amelia wouldn't let him: the rent would eke out their income and the property was something to leave to the family. The two children who are living on the plot pay no rent. One is unemployed and the other recently got married and is saving up to buy furniture for his own house; neither can spare anything at the moment.

## 7  Agustín Yáñez, Guadalajara

Antonio Hernández is 92 years old, a widower who lives with his 62-year old daughter, her painter husband and one of their daughters (who is a social worker). Another daughter and her family come and go, the husband working in different parts of the country for PEMEX. Antonio has five other children. He broke his leg about four years ago and is largely confined to the house, although he likes to collect the rent himself. It does not come to a great deal of money, but he is pleased with the little he gets.

He used to work in agriculture in a rural area of Jalisco. He came to the city about thirty years ago and rented a room in a *vecindad* in this same street. About twelve years ago he bought an empty plot, in instalments, and took about a year to build the house. His daughter and her family came to live there later. Five years later, Antonio started to rent out rooms. He normally has two tenants, keeping one room for his other daughter. He has had some bad luck with tenants not paying the rent. Although he can be bad-tempered he is not very good at dealing with those who don't pay: 'What can you do? They're poor people, just like you'. Eventually, he plans to leave his property to his six children.

## 8  Veinte de Noviembre, Puebla

Manuel Salazar is a former president of the local residents' association. It is said that he used his post to make a lot of money and he certainly lives in a rather attractive house with a garden. His explanation is that, like several other people in the area, he was able to buy his plot with the help of the electricity workers' union; he was an employee of the generating board. He is about 60 years old and married. One of his two children lives on the same plot with his own family. Manuel and his wife live on his pension, the interest from a lump-sum he received when he retired, plus the rent. He claims that the rent, about three-quarters of a minimum salary, constitutes a minor part of his income. His four tenants live next door, in a property with a separate street entrance. Manuel built both buildings with the help of hired labour. Each flat has two bedrooms, a dining room/kitchen and its own services. He started to rent in 1966, and has always had four tenants. Manuel claims that you couldn't rent similar accommodation elsewhere so cheaply. He does all the repairs himself, but even so renting is not a good business. Nevertheless, it is still better to let the property than to have it standing empty.

# 8 Landlord–tenant relations

Commenting on the extensive literature on housing in nineteenth-century Britain, Englander (1983: xvii) complains that 'the relation of landlord and tenant has been ignored by most scholars'. That comment is even more apt when applied to the literature on rental housing in less-developed countries. In this chapter, we examine the nature of relations between landlords and tenants in Guadalajara and Puebla. We consider the average length of stay, the processes by which tenants can be evicted, the methods of selection employed by the landlords and the main sources of dispute between owner and occupier. Finally, we discuss the political organisation of landlords and tenants in the two cities.

## LENGTH OF TENANCY

Security of tenure is rightly considered to be a critical ingredient in landlord–tenant relations. Landlords usually complain of the difficulty of ejecting tenants; tenants, of their insecurity of tenure. We will consider the evidence on eviction below, but first it is useful to establish how long tenants in Guadalajara and Puebla remain in their rental accommodation.

Before considering the figures, it is necessary to establish some yardstick of 'normal' length of tenancy. Unfortunately, this is made difficult by the limited amount of comparative information. What we do know is that tenants in low-income settlements in Bogotá, Bucaramanga and Santa Cruz have relatively short tenancies, averaging between 1.9 and 2.5 years (Gilbert, 1983; Edwards, 1982; Green, 1988a). In Caracas, however, movement in the consolidated self-help settlements is limited and some tenants remain for twenty years or more. While tenancies are more fluid in the central city, with some tenants only staying a few months, many have lived for fifteen years or more in the same house (CEU, 1989: 19–20). Similarly, in Guatemala City, 60 per cent of tenants had lived for more than five years in the same rooms (Rodas and Sugranyes, 1988: 8). Even greater stability is found in central Mexico City where rent control has discouraged many tenants from moving. In the Guerrero district, 60 per cent of one sample had lived for more than ten years in the same *vecindad* and as many as half had lived there for more than twenty years (CENVI, 1986: 180). Similar levels of residential stability have been found in other surveys in the central area of that city; in one set of *vecindades* the average tenancy was eighteen years; in another, it was twenty-two years (CENVI, 1989).

The information from Guadalajara and Puebla supports the idea that most tenants live a long time in the same house. The households we interviewed averaged eight years in their current home (Table 8.1). Admittedly, the length of residence in their previous homes was only four years, but this is probably explained by their moving into better accommodation as soon as they were able.[1] Once they obtain satisfactory rental accommodation, they appear to stay for a long time.

The impression of long stays is confirmed by Table 8.2 which shows that, on average, current tenants have only rented two homes. In fact, as many as 37 per cent of tenants interviewed in Guadalajara, and 56 per cent of tenants in Puebla, had only ever rented one house. Even long-established households have rented very few: households formed more than twenty-five years ago in Guadalajara average only 2.9 rented homes, and in Puebla, 1.9 homes.

The prevalence of long tenancies was confirmed in follow-up interviews with the residents of a *vecindad* in the *barrio* of Analco in Puebla: the mean length of residence was fourteen years and only three of the eighteen

*Table 8.1*  Tenants' average length of residence in rented property (years)

| | GUADALAJARA | | | PUEBLA | | |
|---|---|---|---|---|---|---|
| | Mean | | Median | Mean | | Median |
| *Present house:* | | | | | | |
| Older settlement tenants | 4.9 | 3.0 | (96) | 4.1 | 3.0 | (80) |
| Central city tenants | 10.4 | 8.0 | (118) | 11.5 | 9.0 | (114) |
| *All tenants* | 7.9 | 5.0 | (214) | 8.5 | 6.0 | (194) |
| *Last rented house:* | | | | | | |
| All tenants | 4.8 | 3.0 | (163) | 4.0 | 3.0 | (130) |
| *Last but one rented house:* | | | | | | |
| All tenants | 4.2 | 3.0 | (101) | 4.0 | 2.0 | (65) |

*Source:*  Household survey.
*Note:*  Sample size in parenthesis.

*Table 8.2*  Number of homes rented and length of residential history

| Length of residential history | *Number of homes ever rented (mean)* | | | | | |
| | GUADALAJARA | | | PUEBLA | | |
| | Older settlement tenants | Central city tenants | All | Older settlement tenants | Central city tenants | All |
|---|---|---|---|---|---|---|
| Up to 5 years | 1.4 | 1.6 | 1.5 | 1.4 | 1.5 | 1.4 |
| 5–10 years | 2.1 | 2.0 | 2.0 | 2.1 | 1.3 | 1.7 |
| 10–15 years | 2.5 | 2.4 | 2.5 | 2.6 | 1.9 | 2.2 |
| 15–20 years | 1.9 | 1.8 | 1.8 | 2.3* | 1.6 | 1.7 |
| 20–25 years | 3.0 | 1.4 | 2.1 | 2.5* | 2.4 | 2.4 |
| Over 25 years | 4.0* | 2.7 | 2.9 | 4.5* | 1.4 | 1.9 |
| All | 2.2 | 2.1 | 2.1 | 2.2 | 1.6 | 1.8 |
| Sample size | 79 | 93 | 172 | 69 | 99 | 168 |

*Source:*  Household survey.

*Notes:*  * – Less than five cases.
Many households will also have shared accommodation or had some other kind of tenure during their residential history. For definition of length of residential history, see Table 6.5.

households had arrived during the last ten years; one family had been resident for forty-three years. Ten other households, interviewed in another property in the same street, had occupied their flats for an average of thirteen years.[2] In central Guadalajara, one particularly large *vecindad* contained many long-established households; seventeen randomly-selected tenants had lived there for an average of twenty-one years.[3] Our conversations with landlords confirm that these are not unique cases. In Analco, one landlord had let rooms to several of his current tenants for between fifteen and twenty-five years. A landlady who had inherited her property from her mother stated that some of her tenants had lived there since the flats were built in the early 1950s. An extreme case was found in two properties in Analco in which the resident landlords had grown up with their respective tenants; the tenants' parents had rented a room from the landlords' parents when the current occupants were still children.

It is important to note, however, that there is a clear difference between the central city and the consolidated self-help settlements. The mean length of residence of the tenants in the central areas of both cities was ten to eleven years; that in the more peripheral settlements four to five years (Tables 8.1 and 8.3). In large part this is due to the age difference between the two sets of tenants; the older self-help settlements contain a higher

Table 8.3   Tenants' residential history and length of stay in present house

| Length of residential history | Mean length of stay in present house (years) | | | | | |
| | GUADALAJARA | | | PUEBLA | | |
| | Older settlement tenants | Central city tenants | All | Older settlement tenants | Central city tenants | All |
|---|---|---|---|---|---|---|
| Up to 5 years | 1.4 | 1.9 | 1.6 | 1.6 | 1.7 | 1.6 |
| 5–10 years | 3.0 | 4.2 | 3.5 | 3.1 | 6.3 | 4.6 |
| 10–15 years | 7.1 | 4.6 | 5.9 | 5.8 | 8.1 | 7.2 |
| 15–20 years | 9.6 | 11.6 | 11.0 | 9.0* | 11.7 | 11.0 |
| 20–25 years | 6.3 | 13.8 | 10.3 | 9.5* | 12.3 | 11.7 |
| Over 25 years | 16.3* | 21.0 | 20.2 | 9.3* | 29.5 | 26.0 |
| All | 5.0 | 9.8 | 7.6 | 4.2 | 11.6 | 8.6 |
| Sample size | 79 | 93 | 172 | 69 | 99 | 168 |

Source: Household survey.

Notes:   See Table 8.2.

proportion of younger tenants with much shorter household histories.[4] Interviews with landlords in the older self-help settlements confirmed the impression that long tenancies are more typically a feature of the central areas.

## EVICTION OF TENANTS

Why is the average length of residence in Guadalajara and Puebla so long? There are, of course, several possible explanations but in the light of experience elsewhere, the ease with which landlords can evict tenants would seem to be an important influence. It can be no coincidence that the previously high rates of mobility in England fell markedly when rent controls were introduced at the end of the First World War (Foster, 1979).

In fact, eviction in Guadalajara and Puebla is not an uncommon event. Table 8.4 shows that eviction was responsible for nearly half of recent moves in Guadalajara; in Puebla evictions accounted for around one-fifth of recent moves.[5] While some landlords claimed that they had never evicted a tenant, a majority had tried to do so.

It is clear, however, that eviction can be a difficult business. The fact that some landlords could report cases of tenants having failed to pay the rent for a number of years supports this view. Several landlords also complained that the law was on the tenants' side, but, given that this is the standard

*Table 8.4* Reasons for tenants having left their previous rented homes

| Reason for leaving (per cent) | GUADALAJARA | | PUEBLA | |
| --- | --- | --- | --- | --- |
| | | Penultimate | | Penultimate |
| | Last home | home | Last home | home |
| *Eviction*: | 47 | 45 | 22 | 13 |
| – House to be sold/ demolished/repaired | 21 | 20 | 15 | 9 |
| – House wanted by landlord for own use | 10 | 9 | 4 | 0 |
| – Other/no reason | 16 | 16 | 3 | 4 |
| *Left on own account*: | 53 | 55 | 78 | 87 |
| – To move to better/ larger house/nicer area | 22 | 28 | 46 | 50 |
| – To be nearer work/relatives | 6 | 3 | 7 | 9 |
| – Rent increase/to move to cheaper house | 10 | 8 | 9 | 11 |
| – Personal problems with neighbours/landlord | 3 | 9 | 8 | 13 |
| – Other | 13 | 9 | 9 | 4 |
| Total | 100 | 100 | 100 | 100 |
| Sample size | 144 | 80 | 104 | 46 |

*Source:* Household survey.

complaint of landlord organisations, it was surprising how few made this point.

The legal dispositions concerning landlord–tenant relations in the two cities are to be found in the *Código Civil* of the respective states. In both states, the law requires the use of a contract, which gives both parties certain rights and obligations. Landlords can seek the eviction of tenants who fail to pay the rent: in Puebla, for three consecutive months; in Jalisco, for one month.[6] They can also disregard the contract and seek eviction if the tenant uses the property for purposes other than those agreed. In addition, landlords in Jalisco are able to evict, at the end of the current contract, if they wish to occupy the property themselves or carry out major repairs.[7] Otherwise, providing the tenant respects the conditions of a contract, landlords cannot evict tenants for at least one year in Guadalajara and three years in Puebla even after the contract has ended – a *prórroga*.[8] However, the landlord in Guadalajara is permitted to present prospective tenants with a contract in which they give up their right to a *prórroga*, and this is a practice followed by most landlords. In Puebla, such arrangements are not legally valid, but in fact a standard contract bought from a stationers' contains the stipulation that, should the landlord want the property for any purpose, the tenant promises to leave within fifteen days.

Perhaps the key point about the law on landlord–tenant relations is that it is only helpful if the two sides know what it says. Since many tenants, and, even, landlords do not know their precise rights under the law, actual practice is rather different from the letter of the law. Tenants who sign a contract saying that they may be asked to leave within a fortnight may comply with this stipulation as a result of not knowing that the landlord cannot legally enforce such a clause. Since landlords, on the whole, make greater professional use of lawyers than do tenants, they would seem to be in an advantageous position. Going to a lawyer, however, is by no means a panacea; legal action is inconvenient, expensive, and often a lengthy process. Several landlords, in both cities, reported delays of a couple of years or more in getting a legal case finally resolved. Rental administrators in Puebla claimed that the period was usually three years, and a judge interviewed in Guadalajara, taking a file off his desk at random, showed that it too was already three years old. A representative of the main landlords' organisation in Guadalajara said that court cases could take up to six years.[9] Investigation of some dozens of court cases in both cities shows that, if landlords persist, they will almost certainly win a case. Rental administrators in Puebla put the point more categorically: the landlords, they said, would always win.

Despite winning a case the landlord may still be severely inconvenienced. For example, the tenant may leave without making up the missed rent payments, or, at best, paying the same rent as applied at the beginning of the case. On top of this, the landlord may have legal expenses to pay, although the courts occasionally order seizure of a tenants' goods in order to cover some of the landlord's costs. On the whole, therefore, landlords gain little for their pains in taking the case to court, other than the eventual departure of the tenant. This is the crux of the matter, for the likely support of the courts is of limited value for aggrieved landlords, given the expense involved. In any case, the tenants may manage to live rent-free for long periods of time while they are resisting the owner's attempts to get them removed. A few tenants deliberately exploit their legal rights.[10] The most common method by which a tenant with a grievance against the landlord can gain the protection of the courts is by paying the rent directly to them (*consignación de rentas*). A number of tenants interviewed in the central area of each city were depositing their rent in this way.[11] Depositing the rent with the courts prevents landlords from using their legal right to evict tenants for non-payment; it seemed to be a highly successful policy, apparently feared by many landlords.

Overall, a majority of landlords complained of the difficulties of getting tenants evicted. The strategies they have evolved to deal with this vary. Many court cases end in an agreement between the two parties leading to the tenant's departure, typically after a further three month's residence, rent free.[12] Some landlords may simply condone the existing arrears so that the tenant departs virtually at once; but it is not unknown for landlords to pay tenants a sum of money to guarantee removal.[13] In the light of these fairly extreme measures, it is not perhaps surprising that many landlords try to avoid going to court in the first place: if they are going to end up paying the tenant to leave, they might as well do so straight away and save themselves the extra expense and inconvenience of a court hearing. Some argue with the tenant until the tenant gets tired of the conflict and leaves; others try to handle them 'with kid gloves' and give 'thanks to God when they finally go'. Such an approach is far from that of the stereotyped exploitative landlord. A feeling of resignation about the matter is not uncommon among landlords: 'What can we do about it? Not a lot'.

However, some landlords are not above taking more drastic measures. One landlady in Central Camionera, Guadalajara, was reported to throw buckets of water over families she wanted to leave; others employ 'lawyers' to frighten the tenants. Other tactics include cutting off the water supply to the property, changing locks, or, in extreme cases, destroying parts of the roof, breaking windows, etc.; a few owners and administrators were certainly not above using 'rough treatment' in their efforts to evict tenants.[14] A representative from a tenants' organisation in Guadalajara reported that the tenants' possessions would sometimes be thrown into the street by the landlord. One tenant in Agustín Yáñez said that her family had been evicted from their previous house, because the owners had falsely said that they wanted to use it themselves. They had been offered a sum of money (ten times their current rent) if they left. Since they were not in arrears with the rent and did not want to go, they had refused; the landlord's response had been to send someone round to threaten them.

The overall picture that emerges, therefore, is that the law does give protection to both the landlord and the tenant. However, the main legal barrier to eviction is the way the law operates. The logistics of evicting a tenant are such that it is certainly not easy to get rid of 'difficult' tenants quickly. This is an important reason for the relatively long tenancies which seem to characterise both cities. On the principle of 'better the devil you know', landlords seem to be content to allow any reasonably acceptable tenants to remain in their property for long periods. Replacing them constitutes a risk as an incoming household could turn out to be full of troublemakers. The same reasoning probably applies for tenants: if you are reasonably content with one house and landlord, why risk worse? Moving house almost certainly involves a higher rent, since most landlords charge new tenants more than existing tenants. Troublesome landlords may be a minority, but why take the risk of moving if you can stay?

## LANDLORD–TENANT RELATIONS

Long tenancies are not inconsistent with the impression we received that landlord–tenant relations in the two cities are not generally conflictive. This view is partially supported by the lack of legal disputes under way between tenants and landlords. Fewer than 10 per cent of all the tenants interviewed reported any legal problems in their present house. Most of those lived in the largest *vecindad* studied in central Guadalajara, where the tenants were engaged in a protracted legal struggle with the heirs of the former owners (see below). When asked about any legal problems in a previous house, very few tenants reported any. In contrast, almost half of the landlords reported having made use of a lawyer at some point in the past. Landlord behaviour in this respect, however, varied considerably. Some would automatically go to a lawyer as soon as a tenant started to get into arrears with the rent; others let the matter ride for some time before eventually, and in some cases unwillingly, consulting a lawyer. The length of time before they took action varied considerably; one landlady approached a lawyer after two weeks' non-payment, but several landlords mentioned periods of five to seven months. One gave an initial warning after two months, following it up later, if the tenant had still not paid, by actually consulting a lawyer. Some landlords took legal advice only in cases of exceptional difficulty: for example, when a tenant persistently refused to leave, or was being personally aggressive. Still others never used a lawyer at all – as one landlord said: 'Why should I, when they are poor people, just like me?'

Indeed, an attitude of resignation seemed characteristic of both landlords and tenants. A few tenants even spoke well of their landlords, especially those who had been living in their property for some years. One elderly tenant living alone in central Guadalajara recalled how the mother of her present landlady had rented rooms to single women largely for the sake of their company. She did not charge them very much rent, and in fact she didn't mind if they sometimes forgot to pay the rent. In Agustín Yáñez, a family spoke warmly of their landlord, who had inherited a property from his mother, and treated them well because he wanted them to take good care of his old family home.

The idea that landlord–tenant relations are relatively benign is supported by Table 8.4, which shows that the great majority of tenants in Puebla, and most of the tenants in Guadalajara, had moved on their own account. Few cited problems with the landlord as the principal cause of departure.

Of course, there are many potential sources of conflict between landlords and tenants but the non-payment of rent is the major reason why landlords say they evict. In Guadalajara, one landlady complained of a tenant who had failed to pay the rent for the best part of a year; in the centre of Puebla, two landlords mentioned tenants who had remained in their property for six years without paying rent. Most landlords reported such problems, even if they had not actually evicted the tenants concerned. The second most common reason given for evicting tenants was their drunken or generally unacceptable behaviour; this difficulty was also mentioned by several women tenants, who said that their family had previously been forced to move because of their husbands' drink problem. A third reason for eviction was much less common: misuse of the property. A few landlords had evicted tenants for using the property for commercial purposes; Gabriel Zarate, for example, had evicted a tenant from his house because he was using it as a billiard hall![15]

Whether, of course, the apparently low level of conflict between landlords and tenants is an outcome of the long average tenancy, or the length of tenure an outcome of the harmonious relations, is a moot point. We have no clear evidence on this point. Nevertheless, the general picture of landlord–tenant relations is much more agreeable than that portrayed for Victorian cities in Britain (Englander, 1983; Morgan and Daunton, 1983). It is, however, compatible with new research emerging for a handful of other Latin American cities. In Santa Cruz, relationships between landlords and tenants are generally easy and sometimes genuinely friendly (Green, 1988a: 225). And, in Caracas, 'what most characterises relations between owners and tenants, especially in the self-help settlements, are the informal legal arrangements, the "solidarity", and the mutual trust, which contribute to the "good" relationship, "without problems"' (CEU, 1989: 25).

## THE SELECTION OF TENANTS

Infrequent tenant moves may be due to the difficulty that landlords face in evicting tenants or to the fact that relations between landlord and tenant are relatively peaceful. The latter possibility is more likely if landlords are careful in their selection of tenants. If landlords choose their tenants wisely, or at least exclude those tenants whom they expect to cause problems, this ought to improve landlord–tenant relations.

That landlords are wont to turn away particular kinds of tenant is well-documented in the housing literature. Englander (1983: 10) reports that nineteenth-century British 'landlords of superior working-class dwellings looked askance at applicants from poorer quarters and all took a jaundiced view of their offspring whom they considered the very worst despoilers of property'. Families with children also seem to be unpopular in other parts of the world; they are certainly disliked by Bolivian landlords who prefer quiet and unobtrusive tenants (Green, 1988a). Some landlords have other likes and dislikes. In Karachi, it seems that landlords are very suspicious of letting to single men, who may abscond without paying the rent (Wahab, 1984). Similarly, there is plentiful evidence of ethnic and racial factors influencing tenant choice. In Britain, West Indian immigrants long suffered discrimination at the hands of landlords; in Lagos, in contrast, 'private owners prefer to rent to members of ethnic groups other than their own because ... it is easier to collect rents from those to whom one is not close' (Barnes, 1982: 16).

Whether landlords are able to choose their tenants, however, depends greatly on whether adequate accommodation is widely available; it is difficult for landlords to be fussy if there is a relative shortage of tenants. As Kemp (1987: 13) has recognised with reference to Victorian Britain: 'the level of empty dwellings provided an approximate barometer of the balance of advantage at the margin between landlord and tenant in the urban housing market'. Similarly, it is not always easy for landlords to distinguish between potentially 'desirable' and 'undesirable' tenants. Households with lots of children may be turned away, but what if prospective tenants lie about the size of their families? Written references may be demanded of the tenant, but those references may be worthless. In nineteenth-century Britain, for example, the most common practice was to demand to see the tenant's previous rent book. But what if certain landlords who, 'wishing to get rid of tenants from whom they have failed to extract any rent, and with a view to persuading them to go without giving any trouble, offer to enter up in their rent books all the rent as having been regularly paid when due' (Englander, 1983: 49)?

Our evidence from Guadalajara and Puebla suggests that landlords ought to be in a position to select the kinds of tenants they prefer. There appear to be few empty rooms and landlords report a constant flow of tenants wanting accommodation. One Analco landlord with about thirty tenants claimed that he would see five applicants for every vacancy. Similarly, the great majority of landlords reported that they did not need to advertise empty accommodation. The most that landlords needed to do was hang a notice outside the property, although even this was rarely necessary. Tenants would come of their own accord, asking if there were free rooms or flats. This was confirmed by tenants, only 3 per cent of whom had seen an advert in a newspaper. The majority had heard about the accommodation 'through the grapevine': 41 per cent had known that the property was to let through prior acquaintance with the owner, and a further 39 per cent had either known of a landlord with property to rent through a mutual acquaintance, or been told about the property by a friend or relative living nearby. Most of the rest had found out about their present house either as a result of 'asking around' in the area, or seeing a notice on the door, or because they already lived near by.[16]

Landlords in the two cities certainly express some very clear likes and dislikes about tenants. Some landlords believe that people known to them personally will make the best tenants. Others think that acquaintance inhibits a commercial relationship, so that a measure of social distance is desirable from the landlord's point of view. Most landlords reported that tenants were unknown to them before they moved in.[17] Few landlords admitted to renting to relatives, and the survey of tenants recorded only 4 per cent in Guadalajara, and 10 per cent in Puebla, who were related to the landlord (Table 8.5). Around two-fifths of the tenants reported that they already knew the owner at the time they moved in; a further one-fifth knew of the owner through a mutual friend or acquaintance. This finding is compatible with the preference of a number of landlords for people recommended to them by existing tenants. As a result, it was not uncommon to find a number of interrelated tenant households in a *vecindad*.[18]

The landlords' firmest dislike is for large families; almost all the landlords interviewed would prefer childless couples, or couples with a maximum of two children. Some landlords absolutely refused to take larger families. They fear that children cause damage to their property, and they also worry about the nuisance that children cause, either to the landlords themselves, or to other tenants. The dislike of large families is so deep-rooted a landlord trait that the Civil Code in Puebla tries to legislate against it; contracts forbidding minors to live in the property are illegal. In practice, tenants frequently lie about the number of children they have. Once they are installed in the property, and the contract has been signed, it is too late for landlords to do much about it.

Many landlords also refuse to accept households with animals, because they damage the property or cause a nuisance. One tenant in an Analco *vecindad* complained bitterly that the new landlord would not let her keep rabbits, although her house had its own small patio in which she could keep them, well out of the way of the other households. Many landlords also express dislike for households consisting of single (particularly young) men or women; single people, they argue, are the most likely to be rowdy, or to engage in drunken or 'immoral' behaviour. In practice, there is a limit to the landlords' ability to exclude these different types of tenant. As one old couple renting rooms next to their house in Agustín Yáñez said: 'When they

*Table 8.5*    Prior acquaintance of landlord and tenant (per cent)

| | GUADALAJARA | | | PUEBLA | | |
| --- | --- | --- | --- | --- | --- | --- |
| | Older settlement tenants | Central city tenants | All | Older settlement tenants | Central city tenants | All |
| Owner is a relative | 8 | 1 | 4 | 18 | 5 | 10 |
| Owner was a friend/ acquaintance* | 21 | 24 | 23 | 18 | 25 | 22 |
| Friend/acquaintance in common | 11 | 24 | 18 | 21 | 20 | 20 |
| No direct or indirect knowledge of owner previously** | 60 | 51 | 55 | 44 | 51 | 48 |
| Total | 100 | 100 | 100 | 100 | 100 | 100 |
| Sample size | 95 | 117 | 212 | 80 | 111 | 191 |

*Source:*  Household survey.

*Notes:*   *   Owner was a friend/acquaintance – includes cases in which the landlord was
the same as in the household's previous accommodation.
       **  Excludes cases in which the landlord was not known personally to the
tenant, but in which they had a friend or acquaintance in common (shown
separately above).

arrive, they're as good as lambs – but the majority are almost like animals'.
Few landlords expressed their opinion of tenants in such frank or coarse
terms, but the irritation at prospective tenants giving false impressions of
their situation was widespread.

The main safeguard for landlords obviously lies in checking the tenants
carefully before they move in. In practice, landlords vary considerably in
the care with which they do this. Some simply talk to them, one reporting
that he merely 'gave them a look-over to see what they were like'. Others
take considerably more care. Many take measures to protect themselves
against the non-payment of rent. A common measure is to ask that the
tenant pay a deposit before entering the property. Only a handful of
landlords said that they did not charge one month's rent as a deposit, plus
one month's rent in advance; the great majority of tenants also reported
paying such a deposit. Many landlords quizzed prospective tenants about
their employment, to see if they had a dependable income high enough for

them to pay the rent, and a few asked to see receipts for rent payments from their current place of residence, to check that they were not in arrears.[19]

A more formal method of avoiding problems of non-payment, however, was for landlords to ask tenants to name a guarantor (*fiador*).[20] The majority of landlords asked prospective tenants to name someone, preferably a property-owner, who would pay the rent should they get into arrears. Many tenants are unable to name someone owning property and simply name a relative or friend as guarantor. This is effective when landlords fail to check the guarantor's credentials, but in any case, there can be no guarantee that a guarantor will actually pay the rent owed. In practice, therefore, many of the measures by which landlords try to protect themselves are ineffective. Tenants may initially pay their rent in advance, but they soon drift into arrears. In all but one of the settlements studied between one-half and two-thirds of the tenants acknowledged that they paid the rent at the end of the month rather than at the beginning. A deposit of one month is soon eaten up by tenants getting behind with their payments.

Perhaps the strongest line of defence for landlords is to issue a contract. They are legally required to do so by the Civil Code, and it is in their interests to comply. Without a contract, the landlord is potentially open to abuses such as the tenant making a false claim to the property under the process of *prescripción positiva*. Contracts are usually issued for a period of one year, although a few landlords issue them for shorter periods, usually six or eleven months.[21]

Despite the legal advantages to the landlord of issuing a contract, a significant minority of tenants reported that they had not signed one. In Puebla, one in five tenants had no contract; in Guadalajara, it was two in five.[22] Admittedly, the proportions without contracts in the two cities are much lower than in other Latin American cities. In Bogotá and Bucaramanga, in Colombia, and in Santa Cruz, in Bolivia, contracts are rarely offered to poor tenants, perhaps because landlords can thereby avoid rent controls (Gilbert, 1983; Edwards, 1982; Green, 1988a). In Mexico City, too, only a quarter of tenants in one peripheral settlement had contracts, and only 11 per cent held a legally binding contract (Coulomb, 1981).

Nevertheless, the number of tenants without contracts in Guadalajara and Puebla is still surprising. Sometimes, of course, there is little need for a contract; for example, when the landlord rents to kin or when the tenants have been living in the same accommodation for years and a relationship of mutual trust has developed. In many cases, however, we suspect that the lack of a contract is either due to general ignorance of the law or because the landlord is failing to declare income from rent for tax purposes.

## REPAIR AND DISREPAIR

The quality of rental housing is rarely high and there are consistent grumbles from tenants about their living conditions. Table 8.6 shows that approximately one-third of the tenants in each city reported current problems with the state of their house: leaking roofs, rotting roof timbers, damp walls, walls in need of replastering, locks not working, leaking water pipes, toilets out of order, too few toilets for the number of people using them, dangerous gas facilities, and other problems. Many defects were relatively minor, but some were structural problems which would have entailed considerable expenditure on the part of the landlord.

Approximately one-fifth of tenants in Guadalajara, and one-third of those in Puebla, had complained about a specific repair problem to their landlords. These rather low proportions possibly reflect the wish of many tenants to avoid being seen as 'troublemakers'. The reluctance to complain may also reflect general tenant scepticism about the good it will do them: many landlords simply do nothing. Four-fifths of the tenants who had

*Table 8.6*   Poor housing conditions reported by tenants (per cent)

|  | GUADALAJARA | | | PUEBLA | | |
|---|---|---|---|---|---|---|
|  | Older settlement tenants | Central city tenants | All | Older settlement tenants | Central city tenants | All |
| No problems reported | 67 | 66 | 66 | 73 | 58 | 64 |
| House in need of repairs | 14 | 13 | 13 | 15 | 18 | 17 |
| Services inadequate/in poor condition | 16 | 13 | 14 | 10 | 21 | 16 |
| Other | 4 | 8 | 7 | 3 | 3 | 3 |
| All | 100 | 100 | 100 | 100 | 100 | 100 |
| Percentage complaining to landlord about conditions | 20 | 24 | 22 | 25 | 36 | 31 |
| Sample size | 96 | 120 | 216 | 80 | 115 | 195 |

*Source:*  Household survey.

*Note:*   If more than one type of problem was encountered, tenants were asked to indicate which was the more troublesome.
Sample size is the same for both measures.

complained about the need for repairs reported that the landlord had taken no action about their complaint. Many landlords promised to do something and didn't; others simply tell their tenants to 'like it or lump it'. Landlords can afford to take such a line because the demand for rental accommodation is so high. As a result, most tenants had either to carry out the repairs themselves or to put up with the problem.

In theory, landlords who refuse to carry out the necessary repairs to their property run the risk of being taken to court by the tenant. The Civil Codes in Puebla and Jalisco make the tenant responsible for rectifying minor damage, but major repairs are the landlord's responsibility.[23] Tenants are required to notify landlords of the need for such repairs and, if the repairs are not done, the tenant can take the landlord to court. In Puebla, if the repairs are not done, the tenant is legally empowered to carry out the repairs and deduct the cost from the rent. The courts are also able to award damages to tenants for the inconvenience suffered. Despite this legal provision, we never heard of the courts being used for this purpose.

The impression should not be given, however, that it was only the tenants who complained about the state of repairs: landlords also had a great deal to say on the matter. Indeed, around half of the landlords interviewed counted the damage done to the property by the tenants as their major problem. However, whereas many landlords gave specific examples of tenants who did not pay the rent, few provided details of tenants damaging their property. It seems, therefore, that this is more of a general grumble about the kind of damage and loss to be expected in any residential property, a problem aggravated no doubt by the unwillingness of tenants to put much effort into looking after a rented home. Resident landlords were the most likely to complain about tenants' treatment of the property; they were perhaps the group most directly aware of the deterioration. Typical of such complaints was one woman who said that tenants left the rooms in a 'very dirty' state on moving; another landlady accused her tenants of having broken a tap on the washstand and stolen a gas cylinder.

Landlords also complained about the high cost of repairing and maintaining property. Maintenance is expensive, they argue, relative to the low rents they receive – a view echoed by rental administrators and by municipal officials in Guadalajara in charge of improving *vecindades* in the city. It would seem that this complaint is justified, especially in the case of older property.

In theory, landlords have no choice but to repair the property because the law in both states stipulates that rental property must satisfy certain health and hygiene conditions. In Guadalajara, the municipal Directorate of Public Works even has a small department responsible for inspecting *vecindades*. If inspectors find any infringement of the building and sanitary regulations, they can oblige the landlords to carry out repairs. About 300 interventions of this kind are reported each year, roughly corresponding to the number of *vecindades* classified as being in poor condition, but the fine for non-compliance is very low. In addition, the efficiency of the inspectorate is a little suspect and the Department is not permitted to intervene in the municipal areas of Tlaquepaque and Zapopan. In practice, therefore, control is limited and numerous *vecindades* in both cities are being allowed to fall into disrepair. Indeed, reports of *vecindades* collapsing while still inhabited can sometimes be found in the local newspapers. They are the ultimate testimony to the unwillingness of landlords to maintain their property, and the apparent inability of the combined force of the government and the law to oblige them to do so.

## LANDLORD AND TENANT ORGANISATIONS

Several landlord and tenant organisations operate in the two cities. In Guadalajara, the larger landlords are represented by the Federation of Jalisco Chambers of Urban Property. With 650 members in Guadalajara and representatives of owners in other municipalities of the state, this is an important landlord lobby. It has had considerable success over the years in persuading state legislators to modify legislative proposals.[24] In Puebla, the main lobby is the local branch of the Mexican Association of Property Managers (AMPI), a body which represents the major rental administrators in the city and which lobbies successfully on behalf of landlords. Even tenant representatives admit that the organisation had been very effective in campaigning against radical legislation before 1985.[25] There is also a further small organisation, the Chamber of Puebla Property Owners, which has only 300 or so members, including some landlords. This body is much less influential than AMPI.

Tenants in the two cities are represented by a series of rather weak organisations. In Guadalajara, the Independent Tenants' Union (UII) is probably the most active body. Founded in 1984, it has strong links with the Mexican United Socialist Party (PSUM), which also established the Democratic Tenants' Union (UID) in Puebla in 1982. In the latter city, some tenants are members of the highly active street traders' organisation. The majority party, the PRI, has also established tenant organisations in both cities. In Puebla, the Federation of Urban Settlers and Tenants was founded immediately prior to elections for the state governor in 1986. In Guadalajara, the Revolutionary Federation of Jalisco Tenants was formed in the same year.

It is clear that the main aim of all these associations is to win votes from low-income groups. Their tactics, therefore, tend to reflect the overall strategies of the respective political parties. The PRI organisations do very little, and seek to avoid confrontation with landlords; their main function seems to be to mobilise settlers during election campaigns. In Puebla, officials of the PRI Federation were unable to describe any specific actions they had taken to help tenants; they admitted that the organisation was more concerned with the needs of home-owners in self-help settlements. Similarly, in Guadalajara, the main aim of the PRI tenants' organisation was to acquire plots of land for its members.[26] Nevertheless, it did offer some legal advice to its members and an official in the Federation's office in Central Camionera said that their main contribution was to use party contacts to find new homes for evicted tenants.

The organisations affiliated to left-wing groups were more active and rather more confrontational in approach. Their main role, however, has been to give legal advice to tenants. In Guadalajara, the UII had helped tenants in a *vecindad* in Tlaquepaque who were complaining about the burning of old tyres in a workshop occupying part of the plot. When the landlord retaliated by seeking to evict the tenants, the association defended them; eventually, the landlord gave them alternative accommodation. In fact, the amount of advice given has been limited by the reluctance of tenants to get involved with the associations. Tenants generally do not want to provoke their landlords; they do not want to be threatened with eviction. This fear is aggravated by the possibility that a group of tenants who resist eviction may be charged with stealing the property through squatting, a criminal charge that can land them in jail (Castillo, 1986: 327). This had actually happened to two tenants being advised by a UID lawyer in Puebla. As a result, tenants were seldom prepared to take their problems to the associations. In Guadalajara, the UII reported that they had handled only a couple of dozen court cases and fifty or so *consignaciones de rentas*. In Puebla, a UID lawyer acting on behalf of tenants in three *vecindades* had won six out of fifteen court cases but reported that many of the tenants had simply 'faded away' during the struggle – a common problem.

As a result of tenants' reluctance to pursue court cases, the left-wing tenants' associations have occasionally adopted other strategies. Access to land has been at the forefront of such actions. In Puebla, forty tenant families from Xonaca were among the leaders of an invasion in 1983 which eventually led to the state government supporting the illegal purchase of *ejido* land.[27] In Guadalajara, members of the UII occupied the offices of a government agency to demand plots in a sites-and-services scheme in Tonalá; the state government eventually allocated 200 plots to them.[28] These kinds of tactics have been weakened, however, by an inability to sustain political momentum once the tenants' immediate demands are satisfied, particularly when they have become home-owners.[29]

In the light of this evidence, it is not wholly surprising that few members of tenant organisations were found in the case-study settlements. Ten families who belonged to the UII were interviewed, but they all lived in one *vecindad* in central Guadalajara. They were fighting against the efforts of the late owner's children to evict them.[30] We found no other affiliated tenants. Discussions with the landlords also revealed little sign of tenant organisation. One landlady in Puebla reported that she had been confronted by a tenants' organisation on two occasions, some years previously; but none of the others mentioned any problems.

Similarly, among the landlords there was little sign of political organisation. In fact, none of the forty-seven landlords with whom we spoke belonged to an organisation representing landlord interests.

That so few tenants and landlords in the case-study settlements should belong to active organisations certainly conforms to the history of landlord–tenant relations in the two cities. Over the years, there have been very few documented confrontations either in Guadalajara or in Puebla (Castillo, 1986; Durand, 1984; Regalado, 1987; Estrada Urroz, 1986). Indeed, the main period of confrontation dates back to the 1920s when the anarcho-syndicalists and communists were successfully organising tenants in many parts of the country (Taibo and Vizcaíno, 1984: 166; Durand, 1984: 18-19; García Mundo, 1976). Tenants' unions were formed in both Guadalajara and Puebla at this time.[31] In Guadalajara, the union organised a rent strike, lasting several months, and a number of demonstrations which led to violent clashes with Catholic workers and several deaths (Durand, 1984). The demise of the movement was sealed when the police decided to throw their weight behind the landlords.[32]

Since that time, tenants' movements in Guadalajara have achieved little (Regalado, 1987: 134).[33] Indeed, there have been very few signs of tenant organisation; one member of the UII claims that there was no tenants' union in the city from the 1920s until 1984. In Puebla, there was some activity in the 1940s organised mainly by the Union for the Defence of Tenants (UDI). This organisation had been formed by a left-wing activist in 1940 and was successful in securing a number of minor changes in the regulations governing landlord–tenant relations. It was also partly responsible for the 1948 rent freeze described in Chapter Four (Estrada Urroz, 1986: 150). Increasingly, however, it suffered from repression and 'bureaucratic procrastination' and eventually became defunct (ibid.: 151). Two successor organisations in the 1960s fared little better; certainly, their main aim of achieving a new renting law was thwarted (Castillo, 1986: 294–5).

Even the development of popular urban movements in several parts of Mexico has had less effect in Guadalajara and Puebla than in most other cities. Early meetings of CONAMUP, a confederation of urban organisations in different cities, did not feature representatives from these two cities. The process of mobilisation did lead, of course, to the foundation of the UII and UID in the early 1980s. In Puebla, the UID has links with the traditionally radical Autonomous University, and in both cities members of the radical clergy have been involved in the movements (de la Peña, 1988; Regalado, 1987). One radical priest, in fact, conducted a sociological investigation in the *vecindades* of his parish and reputedly organised a rent strike; he certainly helped tenants fight eviction threats (Marroquín, 1985). His reward was banishment to a parish in another part of the country.

In summary, therefore, tenant associations have been rather ineffective either in mobilising tenants or in influencing the rental legislation. In general, few landlords have felt the need to join associations, their group interests being well represented by organisations such as the Chambers of Urban Property and the Mexican Association of Property Managers. In this respect, the history of landlord–tenant relations mirrors the two cities' generally conservative political traditions.

## CONCLUSION

Tenants in both Guadalajara and Puebla do not move home regularly. The dominant image from the two cities is one of stability of tenure: few tenants have had more than two rental homes and long stays are common. Eviction is not uncommon, occurring mainly because tenants have not paid the rent or the landlords want the property for their own use, but it does not pose a continuous threat to the tenant. Most tenants have annual contracts and the law gives certain rights to the tenant. And, while landlords have superior rights under the law, these rights are difficult to assert. Court procedures are slow, and the determined tenant can stay in the accommodation for a long time, sometimes paying nothing in the way of rent while a legal case lasts. As a consequence, landlords only use the courts on rare occasions. Long tenancies are also linked to the not-especially stormy climate that exists between the majority of landlords and their tenants. While conflict is not uncommon, the average landlord is no more exploitative than the average tenant is a scoundrel. The major sources of problems are not unexpected. The lack of repairs is a perpetual worry for tenants, while poor payment and damage to the property are a source of concern for landlords. The local authorities maintain some kind of vigilance over the physical state of rental accommodation but rarely intervene.

Landlords can choose between tenants because there is a shortage of rental accommodation. They rarely advertise and express a strong preference for some kinds of tenant over others. Households with recommendations from friends and those known by existing tenants are welcomed; families with lots of children or menfolk with a reputation for drinking fare less well in the selection process. Landlords sometimes ask for the name of a guarantor to pay the rent in case of default, although they are generally rather lax in their selection procedure and it is not difficult for a tenant to pull the wool over their eyes.

Few landlords or tenants belong to associations representing their interests. Although such associations are found in both cities, their use of vituperative rhetoric seems to fit uneasily with normal landlord–tenant relations. In general, the larger landlords seem to be better represented than the tenants. There is little tradition of radical social movements in either city.

# 9 The future of renting: policy options

Our study of Guadalajara and Puebla has produced both evidence which supports some existing preconceptions about rental housing and other findings which are much more surprising. It is clear that most Mexican households desire home-ownership, but it is equally clear that not all poor families wish to achieve ownership through a process of self-help construction on the urban periphery. It is clear that many households wish to escape from the insecurity of being tenants, but it is also evident that the picture of the fearsome, grasping landlord is an inaccurate representation of most owners of rental housing. What our results demonstrate is that rental housing accommodates a wide range of different kinds of household and that few overarching generalisations describe it accurately.

While such a finding is not totally surprising, it does pose problems when it comes to formulating policy. The one clear lesson is that it is misguided to think that there can be a single policy which will be appropriate for all forms of rental accommodation. Peripheral rental accommodation does not face the same problems as *vecindades* in the central city; low-income rental housing requires a different approach from high-income housing. Beyond this rather bland, albeit important, point, what policy conclusions can be drawn from this study for both Mexico and for other Latin American countries? The purpose of this chapter is to suggest possible directions in which policy formulation may move.

First, however, we need to address a number of general philosophical issues which complicate the formulation of rental housing policy in any society. Having raised these questions we will return to summarise the lessons of Mexican policy experience and to suggest what the Mexican state might do in the future.

## GENERAL DILEMMAS IN THE FORMULATION OF RENTAL-HOUSING POLICY

Perhaps the fundamental problem in discussing rental-housing policy lies in determining its main objectives. The difficulty is that any policy recommendation touches upon a number of very sensitive political and philosophical issues. Among these issues are: whether renting is a desirable form of residential tenure, whether the state should become a social landlord, and whether financial incentives should be given to private landlords. The resolution of such issues raises wider philosophical questions such as the respective role to be played by the public and private sectors, the necessity and desirability of giving subsidies to the poor, and the extent to which the state should intervene in the process of economic and social change. Because these issues are so politically sensitive their resolution is always likely to be determined largely on the basis of political rationality.[1]

## KEY PHILOSOPHICAL QUESTIONS

### Is ownership superior as a form of tenure to renting or other forms of non-ownership?

The level of home-ownership in different societies shows little connection with any objective criterion of social welfare (Gilbert and Varley, 1988; Chapter Two). The affluent Swiss mainly rent accommodation; the affluent Australians and Norwegians mainly own it; all are reasonably well-housed (Boleat; 1985, UN, 1985). Table 1.1 shows that the 'ideal' of universal home-ownership is close to reality in Bangladesh but the more important ideal of good quality accommodation is rather further from attainment. There seems, in fact, to be no general relationship across countries between levels of economic development and home-ownership. The housing literature reacts to this lack of a relationship mainly in terms of differences in state policy and practice. The impact of the state is most evident in the absence of private home-ownership in many communist countries but it is also a critical ingredient in changes in tenure structure within a single country, as is shown by the dramatic change in Britain from renting to ownership since 1915, and the rise and decline of public housing since 1945. State policy is critical because it establishes the balance of economic and social benefits linked to particular forms of tenure. Thus, ownership is popular in Britain today less because of its inherent advantages than because it is a good investment. It was not so during the nineteenth century for, as Kemp (1987: 11) points out, 'in a stable, relatively inflation-free housing market, the security of tenure provided by occupation leases effectively made owner occupation unnecessary for the well-to-do'. What is desirable in terms of tenure at one relative set of prices may not be desirable at another. Shift the relative prices and either renting or ownership becomes more desirable for the majority. This is the influential role that the state has played; in most societies, it has heavily weighted the balance of economic advantage towards or against ownership. In Britain, tax relief on mortgage payments and rapidly rising house prices make it economically rational to buy a home. In many other parts of Europe, at least until recently, ownership has been less advantageous for the majority. It is difficult, therefore, to distinguish the natural advantages of ownership from the advantages created by the state.

Perhaps, therefore, we should not make too much of the tenure issue *per se*. As Kemp (1982: 4) points out:

> to simply categorise occupiers as either 'owners' or 'tenants' is to ignore an important element of complexity and variation. Hence tenures should rather be viewed as bundles, or configurations, of property rights and obligations, the precise mix of which is liable to some variation albeit within limits. These configurations are not immutable but vary over time and space.

What is accepted as a key ingredient of ownership in one society may be available in rental accommodation in another; the Swiss seem to live contentedly in rental housing.

Nevertheless, it is clear that the needs and desires of different kinds of families are diverse. Within any given society the needs of the young differ from those of the old which are in turn different from those of families with young children. In most societies, therefore, there are groups which are likely to gain more from ownership and others which will find it more convenient to rent. Rental accommodation is particularly suitable for newly-established households, for transitory and mobile groups such as temporary workers or students, and for those who do not wish to tie up their capital in house purchase. Certain households do not want the responsibility of ownership.

Accommodation under different forms of tenure, in different locations, and with a range of prices and quality are an essential prerequisite for satisfactory housing. Unfortunately, not every society offers each group an appropriate kind of housing choice. In so far as governments are able to influence the housing situation, their priority should be to widen the range of choice of land and housing options so that every family can obtain something approaching the kind of housing that they need (see Lemer, 1987).

**The role of the state**

Housing policy is heavily influenced by the ideology and perceived role of the state. Should the state be an arbiter between conflictive private-sector groups? Should it be an economic power in its own right? Should it be a facilitator, easing the path for other groups to create wealth? Decisions made on these issues automatically determine the approach to housing. For example, the state that is hostile to individual ownership will presumably be hostile both to private home-ownership and to private renting. A state hostile to the idea of state intervention will hardly become a social landlord.

Not only the form of intervention, but also its effectiveness, will be determined by the nature of the state. It is an assumption behind most liberal, and indeed socialist, thought that the state is humane and efficient. While that is sometimes the case, it is not the most characteristic feature of the state in many less-developed countries. A major question mark needs to be placed, therefore, against the belief that state intervention will be able to improve the housing situation. Consider a state which is very poor, whose personnel are badly paid and survive more from bribes than from their salaries. Such a state is likely to be inefficient, personalistic, arguably corrupt, and certainly partisan. There is little point recommending sophisticated forms of intervention if the state is incapable of implementing the policy. In some less-developed countries the extension of rules and regulations merely opens up further possibilities for what many would regard as dishonesty and corruption. In devising housing policy, consideration must be given to the ability and willingness of the state apparatus to implement it in the intended spirit.

**Degree to which the general economic/social environment can and should be modified to influence the housing situation**

Running through most of the rental housing literature is an acceptance that housing conditions are an outcome of wider economic and social processes (Harloe, 1985; Kemp, 1987; Daunton, 1987; Howenstine, 1983). National affluence, class divisions, the availability of land, and the organisation of capital markets are among the numerous influences on housing conditions. As a result, national housing policy is hardly autonomous; it is influenced by external factors, such as the world economic situation, the balance of payments and the exchange rate, as well as by internal factors such as the ideological stance of the state, the power of different vested interest groups and the state of the national economy. In short, housing policy is not determined primarily by the needs of the poor or by the collective desire for more housing; it is determined principally by factors exogenous to the housing situation.

In practice, this means that state policy can only influence the housing situation in certain ways, at certain times. It is difficult to build more houses if there is a shortage of bricklayers or cement. It is difficult to interest investors in rental housing if far greater profits are to be made from foreign-exchange speculation. It is difficult to remove rent controls on the eve of an election. Perhaps the constraints posed by the wider economic environment are the critical limitation on housing policy. To a considerable extent, any nation's housing policy is determined more by its Ministry of Finance and the foreign-exchange rate than by the Ministry of Housing.

## Landlords and the unprofitability of rental housing

A critical question facing most governments is: who should run the rental housing stock? Private landlords are one possibility, the public sector (whether local authority or central authority) another, and cooperative/housing associations, another. In many societies rental housing is operated by a mixture of all three. The balance between these different forms is determined, in large part, by the profitability of rental housing. Where housing is profitable the private landlord tends to reign; where it is unprofitable there is a call for the public sector or charities to intervene. In part, therefore, the private/public split is determined by the issue of profitability. In part, it is also determined by the ways in which landlords respond to low profits. If landlords achieve profitability through 'unethical' behaviour, attacking tenants, increasing levels of overcrowding, or failing to maintain their property, the state will be encouraged to intervene.

The problem of unprofitability is resolved in most societies in one of four ways. First, it is resolved by allowing landlords to practise what we have loosely called 'unethical' behaviour. Overcrowding, lack of repairs and services, and regular eviction of poor payers will allow landlords to extract profits. The virtue contained in such unethical behaviour is that at least the poor have a roof over their heads, albeit a roof that probably leaks. Second, the problem can be resolved by encouraging the poor to build their own homes, thereby reducing the need for rental housing. As we have seen, this has been the common response in many less-developed countries. Third, unprofitable rental housing can be maintained, at least in the short term, through the imposition of rent controls. Rent controls are popular with tenants and, since the housing stock decays only slowly, constitute a politically expedient answer to the problem. Fourth, the state can build social housing and itself act as a landlord. The earlier chapters have shown that the Mexican state has followed each of these approaches at one time or another.

There seems to be no way out of this general problem of unprofitability.

It is a problem even in a country as affluent as the United States where the principal problem 'is the increasing paucity of rent paying capacity among the primary consumers of rental housing' (Sternlieb and Hughes, 1981: 12). What is true for the United States is still more applicable to Mexico and other less-developed countries.

## Dealing with poor quality accommodation

Overcrowding, poor services and poor maintenance are the flip side of low rents. In this light what should the state do about bad rental housing conditions? Should poor quality accommodation be left as it is, closed down, or upgraded? In the past, slum demolition has usually been of more help to landlords than to tenants. In nineteenth-century London, urban renewal allowed landlords to redevelop their property; the tenants just moved into other insanitary accommodation (Stedman-Jones, 1971). In general, demolition does not seem to offer much of an answer, especially in less-developed countries. As Abrams (1964: 126) long ago remarked: 'in a housing famine there is nothing that slum clearance can accomplish that cannot be done more efficiently by an earthquake ... Demolition without replacement intensifies overcrowding and increases shelter cost'.

One alternative to demolition, of course, is to make it a statutory duty for landlords to maintain their property. This is certainly the practice in Britain under the 1961 Housing Act (Consumers' Association, 1985: 21–2) and in Mexico a similar kind of proposal was included in the 1983 PRI initiative. It is unlikely to resolve the difficulty, however, in cities where the value of the plot is worth more than the anticipated flow of future payments. In any event, the question arises: what happens to the landlords if they don't maintain the property?

A further alternative is to inject new money into the housing stock. It is clear, however, that if the private sector injects that money, rents will rise. This is hardly a new discovery: in nineteenth-century London 'landlords who exerted themselves to improve their properties ... ejected poor tenants, put up the rents, and attracted another class of occupier' (Stedman-Jones, 1971: 195). If private investment is problematic, then public investment is required. Unfortunately, it is by no means certain that public money will be found to help poorer groups living in rental accommodation. State housing subsidies in most parts of Latin America have often eluded the poor. On the whole, subsidised state housing has gone to the more affluent or to the politically favoured. Since housing subsidies are an important form of political patronage, the 'misuse' of subsidies is difficult to prevent.

In sum, the problem of decaying rental housing is a perpetual one. It is a sad but possibly correct conclusion that 'there is no single one-shot panacea, be it code enforcement, financing, or tax relief, which will substantially improve the maintenance of slum tenements or induce owners to rehabilitate their parcels' (Sternlieb, 1969: xvii).

## Landlord–tenant relations

Is there an inherent contradiction between the interests of landlords and those of tenants? The evidence from a number of societies is that much depends upon the quality of the accommodation involved. Based on observations in several developed countries, for example, Harloe (1985: 288–9) concludes that: 'the vast majority of tenants and landlords co-existed fairly harmoniously. But in the circumstances of decline and lack of resources ... substantial evidence existed of widespread and often bitter conflict'. This suggests that more problems are likely to arise in less-developed countries where most rental housing is in poor condition. In practice, experience is again mixed. Observations in Karachi and Nairobi suggest that there is a great deal of conflict; others, in Caracas, Santa Cruz, Kumasi, and various Indonesian cities, that there is little outright hostility and often a strong measure of landlord-tenant empathy (Amis, 1982; CEU, 1989; Green, 1988a; Nelson, 1988; Tipple, 1988; Wahab, 1984).

Whatever the general climate of landlord-tenant relations, there should be some kind of mechanism for arbitrating in the case of individual conflicts. Arguably, it is the state's role to provide some kind of forum between parties in conflict. Indeed, this is precisely the role that the civil courts play in most countries. Unfortunately, it is quite clear that in most Latin American countries, the courts provide a poor arbitration service. As Urrutia (1987: 59–60) puts it: 'In most countries the judicial systems are overburdened, inefficient and unjust, and few efforts are made to modernise them ... Judges have spectacular case overloads, weak investigative mechanisms, and cases drag on interminably'. As a result, there is little confidence in the legal system. One survey of attitudes in the self-help settlements in Bogotá found that while the inhabitants

> have considerable knowledge of the formal legal system, they do not appear to understand how to use that system, and ... their generally negative or ambivalent opinion of the outcomes under the system suggests that they think that using it would be futile anyway.
>
> (Blaesser, 1981: 130, citing the results of Losada and Gómez, 1976)

In a survey in the *barrios* of Caracas, Karst *et al.* (1973) found that only 3 per cent of households would resort to a court in the event of a dispute over their dwelling.

Lack of confidence in the legal system extends beyond the confines of

Latin America and has prompted the search for alternatives. As the United Nations (1979: 35) notes:

> In most of the countries surveyed ... because the courts (a) are overburdened with general case work and (b) are unfamiliar with the complexities of rent control, special mechanisms have been instituted to implement and enforce rent legislation. Typically, a rent controller's office is set up in the local administrative unit, often with a rent assessment board, committee, tribunal or some similar body assisting the controller and, when vested with judicial powers, acting also as an appelate body.

While that solution sounds thoroughly appropriate, it is less suitable in those countries where administrative procedures are either corrupt or partisan. Again the Latin American experience is not very encouraging. In Lima, for example, Dietz (1980: 63) notes that 'going to the authorities on any level generally produced delays or simply no response; local municipal inspectors ignored conditions or were bought off, and no national agency had the express duty of dealing with central-city rental slum housing'. Admittedly, Pérez Perdomo and Nikken (1982: 224) found that in Caracas, the lowest level of the municipal authority had managed to develop an informal way of dealing with conflicts between neighbours: 'a relatively equitable system of regulations has been established to which there is easy access, in which litigants participate and where decisions are reached quickly'. On the other hand, this informal official system was not without blemish: municipal lawyers arbitrated more on the basis of common sense than in accordance with the letter of the law; worse was that some decisions were made predominantly on the basis of partisan political influence.

One alternative is to leave arbitration to local neighbourhood councils. Since most urban communities have representative bodies, minor disputes might be best resolved by a committee appointed by such bodies. Such was the conclusion drawn by Edwards (1981) and by Karst *et al.* (1973) in their studies of Bucaramanga and Caracas. However, a great deal depends upon the confidence that the local community have in their neighbourhood representatives; in most communities, landlords are likely to have a louder voice than most tenants, and political influence may prove too powerful an ingredient in decision-making (Gilbert and Ward, 1985; Ray, 1969). In this respect, Karst *et al.*'s figures are illustrative: most local people had more confidence in the official bodies than in their neighbourhood associations. In societies where local people have confidence in their representatives, however, arbitration between neighbours should surely be a local responsibility.

**Rent control**

Rent controls have been around for a long time, in Latin America at least since the 1940s (UN, 1979: 2). Current opinion among economists, however, is strongly hostile to controls, mainly because they disrupt the efficient operation of housing markets. Renaud (1987: 187) summarises the criticisms as follows:

> Rent controls can appear to have some short-term beneficial effects particularly for elected politicians, but in the very dynamic environment of TWC [third world country] cities they have three consistent effects: (1) they reduce the rate of return on housing, thereby discouraging investment in shelter; (2) they create arbitrary windfalls and invisible asset transfers from landowners to current tenants in the short run; and (3) they induce illegal side payments in the long run. Rent-control legislation has been shown to be an inferior way to arbitrate conflicts between landlords and tenants: it represents a capricious and unpredictable way of providing subsidies to tenants without appearing to have an impact on public budgets.

However, it is important to remember that rent controls do not work in every society. For a start, landlords are wont to find ways of avoiding the legislation. Frequently, they introduce new methods of charging such as key money (Malpezzi, 1986; United Nations, 1979: 23; Okpala, 1985: 153; Sundaram, 1987: 65). Even where they are effective, there is considerable debate about whether rent controls have been as destructive as many on the right have charged. Certainly, the World Bank's research on rental housing in Bangalore, Cairo, Kumasi, and Rio de Janeiro shows that 'rent control emerges as merely one among several factors discouraging investment in housing' (*Urban Edge*, 1988: 6). Even stronger is the view expressed by UNCHS (1984: 12):

> Comparing statistics concerning private sector production of rental housing prior to and after the imposition of rent control suggests that the impact of rental legislation in this respect may have been marginal *vis-à-vis* a series of other factors, such as the rising prices of land and materials, increasing labour costs and the emergence of other attractive options for capital return, which combined to divert private sector investment towards more lucrative ventures within the housing market.

On the whole, however, the current consensus does favour decontrol, even if most experts agree that existing controls should not be removed too quickly (MacLennan, 1986). If they are dismantled too quickly, the results can be highly damaging for the tenants; in Montevideo, in 1973, 'with housing already scarce and rents artificially depressed for decades, the results of this policy were immediate and dramatic, especially for the poor. Housing quickly became the category of goods to rise most steeply in price' (Benton, 1987: 40). The answer is surely to phase rent controls out gradually, a 'floating up and out' as Malpezzi terms it (*Urban Edge*, 1988: 6).

### The politicisation of housing

This series of dilemmas has been apparent in housing policy in most countries for many years. There are clearly no easy solutions to housing problems; most simple measures create as many difficulties as they solve. In the absence of a simple panacea, therefore, most policy measures are introduced because governments decide to favour one lobby in preference to another. Housing policy is determined largely on the grounds of political rationality. More disturbing is that as a result, policies may be adopted which are both inefficient and inequitable. Informed opinion seems to accept this to be the case with the British housing subsidy system. Despite that agreement, the political situation makes reform impossible. As Daunton (1987: 117) argues:

> It is clear that the subsidies paid in the form of tax relief are inequitable between families, force up house prices, and distort investment. The problem is, of course, that to abolish or reduce the tax benefits would be politically unacceptable.

> In the final analysis it is this kind of problem that fundamentally complicates the formulation of housing policy. It seems to be an unfortunate facet of many societies, that certain reforms cannot be made however desirable or necessary they appear to be.

## SPECIFIC ISSUES ARISING FROM THE MEXICAN EXPERIENCE

So long as a society remains poor, poor quality housing will persist. If the United Kingdom and the United States have failed to resolve their housing problems, countries such as Mexico will certainly fail to solve theirs. Nevertheless, improvements are possible; it is likely that limited changes in the law, the discouragement of certain landlord practices, and the stimulation of tenant organisations may modify and improve rental housing conditions and the general future for rental housing in that country. Some of these efforts are worth examining for their possible use elsewhere; others are warnings that simple measures rarely make for adequate answers.

**The shift from renting to ownership**

A major shift from renting to ownership has occurred in Mexico over the past forty or so years. Many Mexicans would argue that it is desirable for this trend to continue, even though it has been associated with the proliferation of self-help housing. Certainly, the state has no intention of slowing the transition to home-ownership, even if its current policies towards self-help housing could conceivably have that result. It believes that a combination of territorial reserves, well-administered servicing policies, densification, cost recovery and legalisation will produce both more home-owners and better housing conditions. In practice, the outcome may be different. There are many who argue that increasing governmental involvement will damage the prospects of home-ownership for poor families by overregulating the informal or petty commodity sector (Burgess, 1982; Connolly, 1982). Whatever the accuracy of their diagnosis, the transition to self-help ownership in Mexico generally faces a major difficulty compared to the circumstances which favoured such a trend in the 1960s and 1970s. The post-war transition to home-ownership occurred when the country was experiencing a sustained period of economic growth; the 1990s are likely at best to be a period of slow economic growth. Although the poor's share of the Mexican miracle was rather limited, most families experienced rising real incomes and gained fuller access to infrastructure and public services (Bortz, 1984; Ward, 1986). The current problem is that the crisis of 1982 ended that boom and living standards have subsequently fallen dramatically.

Falling real incomes have reduced the chances of many becoming owner-occupiers. Although the cost of land may not have risen during the recession, it is unlikely to have fallen as dramatically as incomes. There is also little evidence that the prices of building materials have fallen. It is possible, of course, that the poor may compensate for rising real costs by occupying small plots, by buying cheaper types of construction material, or by slowing the rate of home construction. If that is the case, the transition to owner-occupation will continue although the size and quality of the accommodation will decline. It is likely, however, that some families will decide that the cost and difficulty of constructing self-help housing is too great. Such families will continue to rent and share accommodation. Certainly, our evidence from Guadalajara and Puebla supports the idea that some tenants in the middle-1980s were too poor to become owners. Our expectation is that the number of such 'enforced tenants' is growing because of the recession.

Even if our pessimistic expectations prove wrong and owner-occupation remains an option for most of the poor, it is by no means obvious that every household wishes to own. Of course, our data showed that most households wanted to become home-owners and that few ever give up ownership to return to shared or rented accommodation. At the same time, we found many cases of 'persistent tenants': families who could have afforded a self-help home and yet had not taken up that option. Among these persistent tenants we found a predominance of the old, the children of tenants, the city-reared, and the centrally employed. The failure to move to peripheral accommodation reflected a shortage of savings, an unwillingness to cope with the rigours of self-help and a preference for life in the centre of the city. The lesson seems to be that, faced by the current barriers to legal, well-located and serviced forms of self-help housing, some households prefer to remain as tenants.

Admittedly, there is a further alternative for these people: the alternative of their buying out the landlords of their existing rental accommodation. Certainly, the Mexican state currently favours such an approach. This is shown by the efforts of agencies such as ISSSTE to sell off state housing to the tenants and by the policy employed after the 1985 earthquake to sell renovated properties to the tenants in the form of condominium ownership. Such a policy is even seen to be a means of remedying the difficulty of dealing with property with frozen rents in the central city areas (Mexico, SEDUE, 1989b). The sale of rental property in condominium ownership is also being actively encouraged by private sector groups such as the Chamber of Construction (Ordoñez, 1989).

This approach points to the fact that, despite its recent advocacy of rental housing, the Mexican state still regards rental housing as a potential nest of vipers. As most Mexicans seem to want to be home-owners, the state wants to satisfy their wish. Any future rental housing policy must be seen in this light.

### Disinvestment in rental housing

While there is a continuing demand for rental housing, it is equally clear that over the years many landlords have withdrawn from the business. Many have converted their property for sale as condominiums; many others wish to do the same. In Guadalajara and Puebla, we met with many landlords who wished to stop renting. Indeed, the majority of landlords interviewed felt that the whole business was unprofitable. In general, we believe that they were right. Rents are low and do not provide a reasonable income; there are better alternatives facing investors, such as putting money into the bank. Managing rental accommodation is also too big a headache for many landlords to face; some were still letting accommodation only because they could not find a buyer. Partly for this reason, organisations such as the National Chamber of Construction are encouraging the idea of condominium conversion. Providing certain changes are made in the law, condominiums development represents a new direction in which to stimulate housing investment. It also constitutes a means by which the occupants can 'live in peace' (Ordoñez, 1989).

Whatever the general consensus among landlords, however, we found some who admitted that renting did offer certain benefits. First, it provided a temporary income from accommodation that might one day house their children and grandchildren. They wanted their families to live with them, or at least near by; rental accommodation could be adapted for that purpose. Linked to this was the feeling that rental accommodation would provide a modest income during the landlord's old age. Since many lacked much in the way of alternatives, rent could form a useful income supplement.

None of this, of course, has a great deal to do with rational business behaviour. Many seemed to admit that beyond investing in bricks and mortar they did not know what to do with their limited savings. A mixture of motives, revolving around the family, old age and a lack of perceived alternatives, seems to characterise the small-scale landlord. None the less, it is this group of the population who are producing the bulk of the new rental accommodation both in Mexico and in other parts of Latin America (Coulomb, 1981; Gilbert, 1983). The irony is that they have done so without any direct help from the state.

Since 1978, the Mexican government has offered incentives to companies building accommodation for rent. Seemingly attractive incentives have been available without stimulating a great deal of construction effort. The effects of inflation, high interest rates, spectacular falls in the value of the *peso*, and (until recently) the enormous returns to be made on the Stock Exchange mean that even generous incentives have met with comparatively little response. Renting's general image has also helped to undermine the policy. For years, landlords have felt that the state has been hostile to their interests. The effects of rent controls, the legal difficulties complicating the removal of tenants, and the day-to-day problems involved in dealing with tenants have made renting an unattractive business option.

However, the 'self-help landlords' continue to invest. If this is the case now, would they not put more resources into construction if they were offered some incentives? Why not extend the availability of current incentives to include this group? Cheap loans or even the offer of cheap building materials might well accelerate the pace of construction. Small sums of money might be very effective in encouraging landlords to invest more heavily in rental accommodation.

Such an incentive programme would face a major difficulty: the fact that so many petty landlords do not have title to their land or currently do not pay taxes. Because of this situation of irregularity, many do not admit to being landlords. Surveys carried out in Mexico City regularly suffer from the fact that both tenants and landlords deny that there is a rental arrangement (Gilbert and Ward, 1985; CENVI, 1989). Landlords not paying income or property taxes would not enter any kind of incentive programme unless they were certain that they could still escape taxation. In fact, perhaps the best financial incentive to the small-scale landlord would be to guarantee immunity from taxation. The other fear among this class of landlord is that in the absence of legal title to their land they may run the risk of losing their property to the tenants as a result of official intervention. It seems that this has been a major worry among peripheral landlords in Mexico City as a result of the recent *Casa Propia* (One's own home) programme. Even though the programme only operates in the central city areas, peripheral landlords are expressing major worries about their legal position.[2] Under these kinds of circumstances, credit programmes to landlords in irregular settlements are not likely to succeed in attracting borrowers.

In addition, the state itself is likely to be hostile to the idea of lending money or giving incentives to landlords in irregular settlements. Given the continued fear that regularisation of tenure is likely to encourage new land invasions and other forms of illegality, the state will be reluctant to encourage renting through this route. However sensible such a strategy may appear, the Mexican state is likely to insist on legalisation first, credit and incentives later.

There is also a further caveat to the attractions of an incentive programme for 'self-help landlords'. Because the landlords who continue to invest in rental housing are not, on the whole, profit maximisers, they may respond unpredictably to financial incentives. If credit is dependent upon their filing loan submissions or submitting planning applications, they may well not respond. If they were more astute and practised investors, there would be no problem; but if they were, they would probably not be investing in rental housing in the first place. Any effort to make them more responsive to economic incentives may simply teach them that there are alternative destinations for their savings. Telling people that they are eligible for low-interest loans may be tempting; on the other hand, it may merely draw their attention to the fact that the main financial institutions are offering investors much more generous returns. Extending the facilities of the formal economy to 'self-help landlords' may simply convince them to put their money in the bank.

What may also encourage further investment by small-scale landlords is the upgrading of the self-help settlements. The evidence suggests that this is a highly effective method of increasing the amount of rental housing, operating both on the demand and on the supply side. Improvements in transport provision and the availability of electricity, water and drainage attracts tenants to a neighbourhood. In Bogotá and Mexico City, upgraded settlements have quickly attracted more tenants (Gilbert and Ward, 1985) and similar conclusions have been reached for many other parts of the world (Keare and Parris, 1982: 9; Robben, 1987: 104; Nientied and van der Linden, 1987). Upgrading appears to be a highly effective method of encouraging rental housing for the poor.

What is less clear is whether upgrading damages the interests of existing tenants. Certainly, there is evidence of rents rising rapidly after service improvements (Moitra and Samajdar, 1987: 77; Nientied and van der Linden, 1987; Keare and Parris, 1982; Robben, 1987; Keles and Kano, 1987). Rents rose both because owners were required to pay for the new services and because the improvement attracted more tenants to the settlements (some drawn from higher-income groups). Whether this led inevitably to the displacement of existing tenants, however, is not clear.

If settlement upgrading has a generally positive effect on the supply of rental housing, so, on occasion, does the establishment of a sites-and-services programme. The lesson seems to be that in many cities the creation of any form of new housing will rapidly increase the supply of rental accommodation. The policy implication is that opening up new self-help areas creates opportunities both for ownership and for renting. Keare and Parris (1982: ix) suggest that this both expands the housing supply and helps owners to repay their mortgages.

## Rent control

Rent controls have a long history in Mexico, but the modern experience really dates back to the country's entry into the Second World War. During the last five years, however, the authorities of several states – Michoacán, Puebla and Querétaro – have approved new forms of control. This legislation reflects both political populism and the concerns expressed by tenants' organisations about the falling real incomes of the poor.

We are less than convinced that the rent control legislation in Mexico actually works. The fact that rents in central Mexico City have risen so rapidly since the introduction of price controls in 1987 suggests that authorised rent rises are being exceeded. While the Ministry of Urban Development explains these rises in terms of the signing of new contracts, which entitles landlords to raise rents, the magnitude of the rises suggests that rents are rising even for tenants with continuing contracts.

Even though we have argued that rents are rather low in Guadalajara and

Puebla and that landlords receive little in the way of rental income, this is not primarily the result of controls. Indeed, rents for property in Puebla, where there is rental legislation, do not seem any lower than rents in Guadalajara, where there is no form of control. The more likely cause of low rents, we believe, has been the combination of difficult eviction procedures, rapid inflation and falling real incomes. It is no coincidence, we believe, that rents have risen broadly in line with rises in the minimum salary.

However, we have too little detailed information to comment fully on this issue. What is evident is that there are numerous problems with the methods used by the Mexican authorities to control rents. First, the criteria used are inconsistent. Some states have linked rent levels to some proportion of the cadastral value, as in Jalisco in the past and, more recently, in Michoacán; others have sought to limit rent rises by linking them to a proportion of the rise in the minimum salary, as in Puebla and the Federal District. Problems arise with both approaches. Tying rents to the cadastral value is sensible providing that realistic property values have been recorded. Since the inefficiency of the cadastral departments in most cities is well known, property values and therefore legally chargeable rents will be low. Linking rent rises to rises in the minimum salary is a reasonable social policy, but does not guarantee either that rents were fair in the first place or that new tenants will pay the same rent as existing tenants. When the minimum salary rises more slowly than prices generally, it also cuts the real value of rents; consequently, landlords may receive too little income to wish to continue letting accommodation. Second, the nature of the administrative bureaucracy in Mexico is such that we doubt whether either method of control has been fairly and effectively enforced.

## Modifying the legislative framework

Laws are often easier to approve than to implement and, in Mexico, the rental legislation is both extensive and thoughtful. The question is whether this legislation has had any real effect. First, is the law obeyed? Second, are the legal requirements understood but generally evaded?

There is certainly evidence that elements of the rental legislation are applied. For example, a majority of tenants with whom we spoke have written contracts, if only because landlords know that the law may favour tenants where there is no contract. Similarly, we found many examples of landlords using the courts, particularly those who made use of rental administrators. We also found a few tenants and tenants' associations taking legal action against landlords or depositing rents with the courts. We also found many examples of landlords following practices that would allow them to sue tenants in the event of subsequent problems: the most obvious example was for the landlord to insist that tenants name a legal guarantor who would be ultimately responsible for paying the rent.

On the other hand, the spirit of the law is not infrequently evaded. The PRI's legislative proposals list many ways in which landlords get round the law (Parcero López, 1983). They mention tactics such as the issuing of non-renewable, eleven-month contracts; tenants having to sign separate credit documents (which not only increases the real cost of renting but opens up the possibility of eviction under mercantile-law procedures); and landlords issuing threats of immediate removal and charging extra payments for services. An important feature of these tactics, of course, is that they are not necessarily illegal: they represent sophisticated responses on the part of landlords to get round the letter of the law. The spirit of the law may be ignored but the law still influences the tenant/landlord relationship.

At the same time, it is easy to recount cases where the law is completely evaded. Many middle-class landlords are so concerned about the difficulty of removing tenants that they let accommodation only to friends and colleagues. No contract is issued and the whole basis for negotiation is one of mutual trust. Letting to friends is some kind of assurance that the property will be cared for and will eventually be vacated. More problematic is that the law is also flouted by many landlords in terms of the non-payment of income tax. Some landlords also demand payment in dollars rather than Mexican pesos.

On the whole, our impression is that landlords are more aware of their legal position than tenants. Whereas tenants talked little about their legal rights, landlords talked extensively about the law. In general, landlords expressed a strong feeling that the law favoured the tenant. They felt especially resentful about the difficulty of evicting tenants who did not pay the rent, who damaged the property, or who behaved badly. This view was expressed even though, in practice, most court cases are won by the landlords. The problem for the landlord is the time that it takes to settle a case. This allows tenants to remain in the property while paying little or no rent. In an inflationary period, with rents held at a low level, speed of eviction is a critical issue.

The general question, therefore, concerns the extent to which the law should influence landlord-tenant relations. Clearly, written contracts should be signed so that both parties are aware of their legal rights. But such rights are worth little if the courts operate so slowly or expensively that many people feel constrained in taking up their legal rights.

Tenure is also a vital issue. Good tenants should certainly have the right to remain in rental property, at least for the length of their contract. Equally clearly, landlords should be able to evict bad tenants quickly and easily. At present, most contracts are issued for a period of one year which seems to be reasonable. There seems, however, to be no reason why contracts should not be signed for shorter or longer periods when the parties agree.

If there is a clear need for legal guidance over the form of written contract this should become part of normal landlord-tenant relations. Thus contracts should be written on the same form, available from any stationers. Use of the courts should rarely be necessary. Indeed, it is our general impression that the majority of landlords and tenants would not use the courts on a regular basis in Guadalajara and Puebla: relations between most landlords and most tenants were not especially difficult. Most landlords did not appear to be particularly grasping; most tenants did try to pay their rents on time. There are hard landlords and some difficult tenants but both are in a minority. The main difficulty is that when there are problems, the courts and the legislative system are too inaccessible and too slow. The consequence is that both landlords and tenants suffer.

The major dilemma is what to do when a tenant cannot pay the rent because of loss of employment or because some emergency, such as a health crisis, has taken all current income. At present, some landlords are prepared to wait a couple of months for payment, but circumstances will arise when this is an insufficient period of grace for the tenant. Clearly, the solution in some kind of welfare state is that the state would provide financial help. In Mexico, however, aid of this kind is not readily available so that the law has to determine whether or not tenants unable to pay rent should be obliged to leave. The form of resolution of this dilemma lies in the determination or otherwise of the state to encourage rental housing. If it wishes to stimulate investment, then it must allow landlords to evict tenants under these circumstances. Any policy that allowed tenants to remain despite the non-payment of rent would undermine landlord confidence.

## More effective landlord and tenant organisations

Too many tenants are ignorant of their legal rights; too many landlords are unable to uphold their rights because of the slowness of the courts. One method of improving understanding among tenants is by providing cheap legal advice through tenant associations. One method of helping small-scale landlords is by encouraging landlord organisations. At present, the majority of landlords and tenants do not belong to representative associations. Those who do belong sometimes benefit; in Guadalajara and Puebla, academics have organised a panel of lawyers to advise tenants. It seems very important that more representative associations be developed for both landlords and tenants. At the very least, these should offer cheap advice about the law and about normal landlord-tenant practice.

Of course, there is always the danger that such associations will develop predominantly into partisan political bodies; indeed, the current associations in Mexico are all highly politicised. This is a danger but less of a problem than the absence of representative associations which guarantees that the rights of neither tenants nor landlords will be served.

## The central city problem

The most obvious accommodation problems are found in the decaying central areas of the cities. The basic dilemma is that market values for the plots are higher than the value of the building and the flow of future rents. Sometimes, too, the rents are worth less than the cost of necessary repairs. As a result, many landlords are allowing their properties to decay: some are choosing not to repair their property; some cannot afford to do otherwise.

Urban renovation has led to the eviction of large numbers of tenants in the central city mainly through road improvement schemes (Coulomb, 1989: 175). Most other tenants have been neglected by the state, even if a few have continued to benefit from the rent controls introduced in the 1940s. In fact, the general neglect of the central city rental housing stock was only reversed in Mexico City as a result of the 1985 earthquakes.

Whatever the longer-term impact of the earthquakes on Mexican housing policy, it demonstrated that there is a continuing need for economic accommodation in central areas. Certainly, our interviews reinforced the conclusion that, in the case of Guadalajara and Puebla, there was a continuing demand for central accommodation. Many households were prepared to tolerate poor environmental conditions rather than move to the periphery of the city. Their decision was eased, of course, by cheap rents, but many stated clearly that location was important. They were prepared to foresake the option of self-help ownership in part because such an option was only available on the fringe of the city. As Mexico's cities grow larger, and the journey to work from the periphery to the centre of the city gets longer, the option of a central location becomes more important.

A possible method of maintaining such central accommodation is for tenants to purchase their current homes. Indeed, this is the current policy of the Mexican government in Mexico City, and, in principle, throughout the Republic. The earthquake relief programme and its follow-up programmes, such as *Casa Propia*, seek to relieve the problem of deteriorated housing by transferring ownership from disgruntled landlords to the existing tenants (Lugo, 1989). The hope, then, is that the new owners will invest their savings in making improvements to the property. The first question is whether all tenants will wish to buy. Certainly, those who have lived a long time in the same accommodation may be tempted, especially if they have been affected by the recent rises in rents. Clearly, most Mexicans do want to be home-owners: a desire which can be satisfied through this kind of programme. Whether, however, tenants living in accommodation with frozen rents will be readily attracted is another question. The government hopes that the attractions of ownership will be sufficient incentive, but we are less sure. It was one thing to buy property under the highly subsidised earthquake relief programme; it is quite another to buy without a subsidy. Second, there is the issue of whether the new owners will in fact be able to afford the cost of improvements. It will be interesting to see what happens with current programmes in Mexico City.

Certainly, there is no question that groups of tenants should have the right to buy their property. Unfortunately, encouraging existing tenants to buy property may well cut off new sources of central accommodation for newcomers. The current policy is no solution for temporary workers, newly-established households, or households with jobs in the central city. Even if the quality of the accommodation improves it will only be available to those with the ability or right to buy.

Of course, the policy to turn existing tenants into owners of condominiums is not being applied outside Mexico City and even in the capital it is not certain of success. As such, the question remains how to upgrade deteriorating rental property; what alternatives are available? One possibility is to adopt some variation of the *patronato* schemes tried in Guadalajara. After all the schemes did put some pressure on landlords to repair their property and led to widespread minor improvements. But Guadalajara is an unusual city in Mexico in so far as local government works tolerably well and there is a great deal of collaboration between the private and public sectors. Calls to civic pride are more likely to work in Guadalajara than elsewhere. In any case, the schemes represented no more than one step in the right direction.

Perhaps the only remaining alternative is to attempt to channel the resources of the tenants into maintaining and improving the property.[3] This is something that the Mexicans have not yet managed to do. One possibility is to divert rents into maintenance: 'In Oporto (Portugal), tenants can undertake essential repairs after 120 days and offset costs against rent payments' (MacLennan, 1986: 23). Another is for tenants to contribute to a fund for rental housing improvements. Coulomb (1985a: 45) recommends that they should contribute 5 per cent of their rent to a fund controlled by the local authority.[4] Clearly, such a fund would increase the financial burden on tenants. Problems would also be involved in managing the fund efficiently and in selecting particular properties for improvement. Nevertheless, it is a means of generating a guaranteed source of funds for property repairs.

## Housing subsidies

Traditionally, housing subsidies in Mexico have gone to owner-occupiers. More often than not, the funds have been devoted to the construction of purpose-built housing. Currently, this is a period of austerity for the Mexican government. Nevertheless, it continues to channel moneys into housing construction; there are even subsidies for building housing for rent. Unfortunately, too little encouragement has been given either to 'self-help landlords' or directly to tenants.

While the subsidies available will never match the need, it would seem sensible to transfer some resources from owner-occupiers to the rental sector. Of course, such a decision will ultimately be made on political grounds, and, so far, the trade unions and employers' associations have successfully channelled state funds into purpose-built housing. Perhaps it is incumbent on the tenant associations to make a stronger plea for subsidies?

## A FINAL REFLECTION

Unfortunately, we have failed to discover a panacea that will remedy past and present ills. Mexican experience has not been very encouraging in this respect, which is perhaps why the country's government is so ambivalent in its attitude to the rental sector. Despite the recent introduction of incentives for investors building rental housing, the main thrust of Mexican policy is to continue to foster home-ownership. Perhaps this is the correct solution, even if the poor may face major difficulties in becoming home-owners. It may even be the most appropriate policy for expanding the rental housing stock. For as we have seen, small landlords are the main source of new rental housing. Wherever self-help housing has been allowed to develop, particularly when peripheral settlements have obtained services and transportation, levels of renting have risen dramatically. While it is undoubtedly ironic, perhaps the best strategy of improving rental housing is to encourage self-help construction. It has always been something of a myth that this route would offer every family the chance of home-ownership, certainly in the shorter term. Perhaps we have finally disproved this myth by demonstrating that it is self-help owners who are the principal producers of rental housing.

What we may also have shown is that the relationship which exists between renting and ownership is symbiotic. Not only is renting a route into ownership, but the consolidation of an owner's home is often financed by the rents of tenants. If this is the case, then we must continue to strive for the introduction of appropriate policies towards self-help housing. What form these appropriate strategies should take is, of course, locally specific and open to debate. One thing, however, is certain. In future, self-help housing programmes cannot afford to ignore the issue of rental housing.

Of course, our failure to produce a panacea for the problems of rental housing is unfortunate and is certainly not the conclusion that we wanted. However, we are hardly the first authors to discover that there are no perfect answers. As Sands (1984: 56) reflected on the basis of his analysis of housing in the United States, 'given the complex nature of the economic, social and demographic factors which impinge on the rental housing market, this inability to provide a simple solution to the rental housing dilemma is perhaps unavoidable'. At least, such an admission avoids the danger of overseas academics telling governments what they should or should not do. We can only hope that we have alerted them to the need for thinking much more deeply about the issue of rental housing.

# Notes

## Preface

1 Like that of Edwards, her research was funded by a grant from the Economic and Social Research Council.
2 This study is using a methodology closely related to the one employed in this study. It is being carried out by Oscar Olinto Camacho, Rene Coulomb and Andrés Necochea, and coordinated by Alan Gilbert.

## 2 Research strategy and a brief guide to Mexico

1 Known elsewhere in Latin America by different names such as *conventillos*, *inquilinatos*, *cortiços*, and *mesones*. While there is no completely satisfactory definition of the term *vecindad*, most are instantly recognisable in practice. The main door from the street opens onto a communal area, sometimes a patio and sometimes a central passageway. Surrounding this communal area, on one or two storeys, are numerous dark, often windowless, rooms. Densities are high, services are communal and usually deficient, and the building is often in poor physical condition.
2 The *ejido* was established under the land reform of 1915. It granted agrarian communities the right to farm land in perpetuity. Members of the community had the right to cultivate land. It was not permitted to sell the land.
3 For our purposes a household was defined as 'an autonomous group of people living under the same roof who normally eat together and share a common budget to cover the costs of accommodation and subsistence'.
4 Owners were defined as having the legal or *de facto* right to occupy, let, use or dispose of the dwelling (and usually the land on which it is built); or they could be in the process of acquiring that right (e.g. through payment by instalments). An owner has no obligation to pay any rent or charge to a landlord. Tenants were defined as households paying a prearranged rent for the exclusive occupation of all or part of a house. The accommodation will normally have a separate entrance from those of other households in the same dwelling or plot. A sub-tenant is a tenant of a tenant.

## 3 Residential tenure in urban Mexico since 1940

1 The Mexican census threshold for an urban place.
2 In 1960, the four largest cities were Mexico City, Monterrey, Guadalajara and Puebla.
3 1970 census data on tenure must be interpreted with caution. Examination of the results for individual enumeration districts in Mexico City suggests that there is considerable confusion between renting and illegal ownership. Similar settlements on *ejido* lands may be recorded as having either 0 per cent or 100 per cent non-owners, presumably in relation to the enumerator's understanding of their legality; such differences clearly have nothing to do with the proportion of tenants to be found in those settlements (Varley, 1985b).
4 It should be noted that in the absence of pre-1950 data on tenure, it is possible that these high proportions of non-owners represent a decline from even higher levels of non-ownership previously.

## 4 Mexican housing policy

1 The earthquake in Mexico City certainly stimulated tenant activity in the city. An important new association, the Asamblea de Barrios, was claimed to have between 20,000 and 30,000 members (*Proceso*, 20 July 1987), and protests spread to other aspects of urban life such as the levels of taxes and service costs (*Proceso*, 16 February 1987).
2 Officially, there were 1,000 families living in the settlement, but some accounts suggest that as many as 500 were living on the 100 hectare site (*Proceso*, 7 November 1988).
3 There may well have been similar efforts by state authorities in other parts of Mexico although we are unaware of any concrete examples.
4 Initially the tax relief was on 4 per cent (and in certain circumstances 6 per cent) of the total value of the finished homes.
5 This kind of housing is known by the acronym VIS-R: Social Interest Housing for Rent. The housing has a minimum constructed area of 45 square metres and consists of a living room, bathroom, kitchen, two bedrooms and a service patio (Mexico, FOVI, 1986: 6). Rents must not rise faster than the minimum salary. Any tenant who has lived in the property for three years, and is up to date with payment of the rent, has the option to buy.
6 For example, in Argentina in 1943, in Brazil with the introduction of 'Modern Rent Law' in 1942, in Chile with the modification of existing laws in 1941, and in Colombia with law 0453 of 1943; similar legislation was introduced by many British colonies slightly earlier (UN, 1979). Clearly, Latin America's entry into the Second World War, and the general inflation that was afflicting the region as a result of the hostilities, precipitated the legislation.
7 After several years of low inflation, prices rose by 10 per cent in 1941–2, 18 per cent in 1942–3, 33 per cent in 1943–4, 6 per cent in 1944–5, and 27 per cent in 1945–6 (Wilkie, 1984). The inflation rate then slowed until 1950–1.
8 As Aaron (1966: 316–17) shows, the basis of these calculations is extremely flimsy.
9 The conversion of rental property into condominiums for sale had become something of an epidemic by this time.
10 The President of the *Gran Comisión del Congreso del Estado* declared that

only the January increases in the minimum salary would count. He made this point when noting the ambiguity about the annual increase.

11 Study being undertaken by Jaime Castillo at the Universidad Autónoma de Puebla.

12 In the State of Michoacán, the legislation stipulates that the rent must be no more than 0.5 per cent of the cadastral value (Coulomb, 1985a: 24–6). The contract is to last at least two years, although tenants can end the contract with two months' notice. A deposit of one month can be charged as a guarantee for the landlord. Guarantors are not required, and no embargo can be placed on goods of the tenant. A fine equivalent to three months' rent can be charged if the landlord charges more than the rent laid down in the contract.

13 In 1959, it was claimed that the city had 160,000 people living in 1,412 *vecindades* with 7,380 rooms (Jalisco, *Informe de Gobierno del Lic. Agustín Yáñez*, 1956 and 1959). A physical improvement scheme was introduced under which 48 landlords had undertaken improvements in 1957, 220 in 1958, and 410 in 1959. Expropriation was threatened if landlords did not cooperate, especially with respect to providing water. Credit was available, although no landlord took up the offer, nor did any sell their property to the local authority. There was also a social element to the campaign with health teams and subsidised foodstuffs. By 1970 it was estimated that there were some 1,600 *vecindades* in the city.

14 In Puebla, FONHAPO provided the tenants of one property in Analco with credit to buy their homes from the owner. The same owner was trying to achieve a similar agreement for another property in the same area, through her administrator. These were the only two projects in hand by late 1986, and other rental administrators had not even heard of the scheme. In Guadalajara, FONHAPO has done little; we know of only one scheme in the central area, affecting twelve apartments. It should be noted that these efforts pre-date the earthquake.

## 5 Urban development and the housing market in Guadalajara and Puebla

1 The sparseness of the Indian population in the Guadalajara region would have implications for the post-Revolutionary agrarian reform, and, consequently, for the development of twentieth-century Guadalajara, as will be seen below; it also contrasted with the much denser Indian population in the Puebla-Tlaxcala region.

2 The new industrialists who emerged in post-Revolutionary Guadalajara took advantage of the Second World War to consolidate their enterprises (Kruijt and Alba, 1988). They included Calzado Canadá, which would become the largest shoe manufacturer in Mexico.

3 The large-scale sector also benefits from the informal sector by extensive use of outwork (Gabayet, Lailson and Padilla, 1987).

4 The reasons for the importance of Puebla's cotton textiles industry were its proximity to the colony's main markets and its location on the trade route from Veracruz – the major cotton producing region until the nineteenth century (with foreign imports, at times of insufficient domestic production, also coming through the port of Veracruz) (Gamboa, 1985).

5 Puebla retained its status as Mexico's second largest city until the 1860s, when it was overtaken by Guadalajara (Kemper and Royce, 1981: 10).

6 References to Census data refer, with the exception of Table 5.3, to the central *municipio*, plus, in the case of Guadalajara, Tlaquepaque from 1950 onwards, and Zapopan from 1960 onwards; and, in the case of 1960 data for Puebla, the following five *municipios*; San Felipe Hueyotlipan, San Jerónimo Caleras, San Miguel Canoa, Resurrección and Totimehuacán. Data for these five *municipios* are included in the Puebla figures for 1960 because they were annexed to the central *municipio* in 1962. Some statistics therefore reflect the incorporation of extra *municipios* in particular years.

7 It has been argued, however, that the predominance of small, often unregistered manufacturing enterprises in Guadalajara leads to considerable under-reporting of industrial employment, and that industry may even be the city's major employer (Escobar, 1986a; Wario, 1984).

8 Even during periods of greater union activity, such as the 1920s, the strongest unions were the 'pro-employer' Catholic ones (Durand: 1984). The traditional importance of the Church and its concern to establish working-class organisations to counteract the left (see de la Peña, 1988), may therefore play a part in explaining low wage rates in the city. However, Escobar (1986b) argues that unionisation makes little difference to salaries in Guadalajara compared with those in other cities.

9 The sectoral composition of each city's industrial work-force obviously affects its overall performance. Examining the record for individual sectors confirms Guadalajara's poor performance; Puebla does not do much better. Guadalajara wages were better than those in the other cities for the shoe industry, and the manufacture and weaving of artificial fibres; but the city does not seem to have performed well in other sectors, and decidedly badly in pulp and paper production or iron and steel. Puebla performed well in the production of car parts and accessories (but not particularly well in vehicle assembly), soft drinks and electricity generation; and badly in textiles (only artificial fibres considered), ceramic goods, and cement. However, comparison with earlier studies in the same series suggest that the situation varied considerably from year to year, as several of Guadalajara's industries did relatively well (better than their Puebla counterparts) in 1979, for example.

10 Growth rates based on Table 5.3. It is widely believed in Mexico that there was a considerable degree of under-reporting in the 1980 Census.

11 Unless stated otherwise, all areas quoted for the two cities are calculated from computer analysis of the maps reproduced in Figures 5.1 and 5.2.

12 However, Walton (1978: 40) observes that the trend described by the Dotsons had in fact been established much earlier, with 'new colonies' such as the 'American' and 'French' colonies developing along the major route to the west from the early 1900s (see also Vázquez, 1985: 60–1).

13 A major extension to the trolley-bus system was inaugurated in 1989.

14 From 31 km$^2$ in 1945, the city grew to 67 km$^2$ by 1955, and 162 km$^2$ in 1972.

15 Puebla grew from 14 km$^2$ to 23 km$^2$ between 1950 and 1965.

16 The limits of the urban area of Puebla are less than clear. In addition to five municipalities absorbed through an administrative reorganisation in 1962, Méndez Sainz (1987) includes a further fourteen nearby municipalities in the city. In contrast, Negrete and Salazar (1986) include only eleven additional administrative areas in their definition, even though some such as San Martín

Texmelucan are more than 21 kilometres from the centre of Puebla.

17 The other most important ecclesiastical centre of colonial times was Mexico City, which had the third highest proportion of rental homes in 1950.

18 It should be noted, however, that another source indicates that although the housing deficit for the State of Jalisco was 63.7 per cent of the existing stock, compared with a national average of 46.1 per cent, the figure for the Metropolitan Area was only 43.5 per cent (Jalisco, DPUEJ, 1979: 72).

19 In Table 5.4, the data for persons per room is an estimate, representing the maximum possible figure, because the final category of the number of rooms in each house is always 'n or more' rooms. In addition, for 1980, the assumption is made that the population living in houses for which the number of rooms is not specified represents the same proportion of the total population as the number of houses-with-unspecified-rooms as a proportion of all houses.

20 Vázquez (1984) gives much credit for Guadalajara's image as a city well provided with services to the high level of public and private collaboration in the city, manifest most notably in the Council for Municipal Collaboration. While this system certainly worked well from 1959 to 1975, it has been less successful since. It has also been criticised for undertaking many unnecessary infrastructure projects, an outcome of its principal function which is to provide work for local construction companies (Wario, 1984: 163–4).

21 If, however, this argument is correct, we might expect that services would apparently get considerably worse in Guadalajara and Puebla at the time they underwent rapid physical growth; given the different timing of that process in the two cities, the changes in servicing standards might be expected to occur at different times. The data in Table 5.4 suggest that servicing levels deteriorated considerably between 1950 and 1960 in *both* cities. However, this apparent deterioration may be partly explained by changes in the Census definition of what constitutes a domestic water supply.

22 Arias (1985) notes that there were no fewer than twelve governors of the state of Jalisco during the 1920s; not until the mid-1930s did a governor manage to last out his six-year period of office – but two succeeding governors only managed to last four years each, so that full political stability was not really achieved, from one point of view, until the late 1940s.

23 The relative scarcity of indigenous villages in the region, noted above, was one factor in Jalisco's relative quiescence during the Revolution and the later application of the agrarian reform than in some parts of central Mexico (Sanderson, 1984).

24 Information from file on the *ejido* of Tlaquepaque in the Jalisco *Centro Regional* of the *Cuerpo Consultivo Agrario*. Vázquez (1989) provides several other examples of landowners subdividing land for housing in an attempt to subvert the reform.

25 The subdivision was originally named after the landowner, before its name was changed (perhaps for tactical reasons?) to that of the Governor of Jalisco from 1953 to 1959. The information about the subdivision is taken from its file in the Municipal Directorate of Public Works Subdivision Office.

26 A list of subdivisions registered with the municipal authorities for tax purposes in 1985 (however old the subdivision) lists fourteen developments by the man responsible for Agustín Yáñez, and four by the family of the woman subdividing her estate near Tlaquepaque in the 1930s; two other individuals or families are registered as the owners of seven developments each, one of five,

and several of two or three. The majority of *fraccionamientos*, however, are listed as belonging to companies, which are likely to represent associations of the same groups of people. Several of the names can be linked with old landowning families of Guadalajara and/or the city's business elite (as listed in sources such as Kruijt and Alba, 1988: 5, and Sánchez, 1979: 82).

27 This man's company was still involved in subdivision in the mid-1980s (Varley, 1989b).

28 Increasingly, from the 1960s onwards, middle-class areas in the west of the city had been developed by large property companies linked to national financial interests; these companies were responsible, not only for subdividing the land, but also for construction and provision of finance for house purchase (Wario, 1984: 158). During the 1970s, the link between financial institutions and housing development largely replaced the old system of individual building projects (*construcción por encargo*), leading to the emergence of new firms (over 70 per cent of building firms in Guadalajara have a post-1970 origin) and increasing domination of the sector by the larger firms – a trend in which government 'social interest' housing finance has played a role (Rodríguez Ortíz, 1984: 167–71). A list of subdivisions prepared by the Guadalajara tax authorities in 1985 lists several dozen developments registered in the name of financial institutions; more recently, it is rumoured that property development has been used to 'launder' money belonging to the drugs barons from the north-west who have made their home in the city. The frenzied investment in the property sector of the 1970s helped drive up land prices (Wario, 1984: 159); the potential profits and greater security of investment in middle-class housing construction helped put the final nail in the coffin of the low income commercial subdivisions.

29 Agrarian Department officials did, however, make some attempts to intervene in the process to stop the proliferation of *ejido* land sales. For example, in 1971 the Department's Chief Officer sent the Directorate of Prosecutions a denunciation by Tlaquepaque *ejidatarios* against illegal land sales, to a value of 200,000 pesos, by their local leader (from file on the *ejido* of Tlaquepaque in the regional offices of the Agrarian Consultative Committee).

30 One of the resulting settlements was given his name.

31 In fact, the CNC group kept the money, threatening *ejidatarios* that they would go to jail if they tried to pursue the matter. Information on Buenos Aires from an interview with a member of a community action group working in this area who had lived in the settlement for a number of years.

32 Information from interview with a North American estate agent/property developer resident in Guadalajara.

33 Presentation by the Director of Urban Land of the State Planning Authority at a seminar in the College of Architects and Engineers of Jalisco, 13/11/85.

34 Information from the Agrarian Reform Ministry.

35 The reasons for this are to be found in the persistence of a high proportion of independent Indian communities in the region: communities which had seen the depradation of their lands by *hacendados* in the nineteenth century, to the extent that 90 per cent of villages and towns had no communal lands at all left by 1910 (Simpson, 1937: 31).

36 The agrarian reform was facilitated by the persistence of independent communities in the region, since resident workers on the *haciendas* (*acasillados*) were excluded from the reform. Consequently, whereas only 23

per cent of communities in Jalisco were eligible for reform, 70 per cent of Puebla villages were legally qualified to benefit from the reform (Sanderson, 1984: 45). Nationally, most land grants took place during the Presidency of Lázaro Cárdenas (1934-40); but by 1935, 78 per cent of all land reform beneficiaries in Puebla had already received their grants (ibid.: 86).

37 *Colonia* El Salvador was founded on *ejido* lands belonging to the village of Chachapa, just outside Puebla to the east. This *ejido* was granted 800 hectares in January 1918, and another 652 hectares in 1931. The contrast with the *ejido* of Tlaquepaque in Guadalajara (which provided the lands on which *colonia* Buenos Aires grew up) confirms the contrast between the two cities: Tlaquepaque was founded in June 1937. The early nature of the reform around Puebla is likely to owe something to the close links between some of the city's textile elite and the Díaz (pre-Revolutionary) and Huerta (counter-Revolutionary) regimes (noted in Gamboa, 1985).

38 Information about Veinte de Noviembre comes from files and a map of the area in the state Treasury (cadastral division); plus an interview with the daughter of one of the intermediaries responsible for selling plots, who still lives in the area. The authorisation of a subdivision of this name is recorded in a list of subdivision authorisations collected by Dr Patrice Mele from municipal and state planning authority files.

39 However, as Mele (1988b) notes, the lack of subdivisions registered prior to 1960 may also reflect an incomplete archival record.

40 Mele (1988b: 15) presents a list of subdivisions and their developers. Over half of the authorisations for 'popular' subdivisions involved areas of less than 4 hectares. Most developers (individuals or companies) appearing in the list are recorded as responsible for only one subdivision (although some developers may have been involved in more than one company).

41 In the mid-1940s, 'urban zones' (areas set aside for *ejidatarios*' houses but invariably occupied by non-*ejidatarios*) were formed in the *ejidos* of Chachapa and San Jerónimo Caleras. Another was founded in San Baltazar Campeche in the 1950s. Information from documents in the Chachapa file in the Puebla Delegation of the Agrarian Reform Ministry.

42 The close involvement of the *ejido* community in El Salvador may be seen in the fact that, although the settlement is some distance from the village of Chachapa, 5 per cent of residents were relatives of Chachapa *ejidatarios*. In contrast, no such people were found in the Guadalajara *ejido* settlement studied.

43 One estimate of the cost of plots on *ejido* land in Totimehuacan on the fringe of Puebla in May 1986 was 1,500 to 2,000 pesos per square metre, another estimate was 1,200 pesos per square metre in San Baltazar in September 1984. An earlier published estimate of the cost of land in popular settlements in 1978 was 300 pesos per square metre (Puebla/SAHOP, 1978: VI Anexo, Plano Microregional). Expressing these prices in terms of the minimum salary of the day gives the cost of a square metre of land in 1978 as the equivalent of three days' minimum salary; in 1986, of between one and two days' salary.

44 Mele (1988b: 16) notes that there are two forms of speculation. The one, open only to the wealthy, is to buy up extensive tracts of land on the fringe of the city. The other is to buy one or several plots in a settlement and wait for the price of land to rise. Both methods have been common and as a result there is a great deal of vacant land in both cities.

45 The data were collected from newspapers for the months of September and October in each year.

## 6 Residential tenure: choice or constraint?

1 Similarly, in Egypt, 'ownership status, despite being preferred by households, is not positively related to income' (Abt Associates Inc, 1982: 125).

2 As noted in Table 6.1, 'sharing' here indicates *either* households sharing a plot with the owners, but not forming part of the owners' household, *or* households living as part of an extended household, in any kind of tenure, without being the householders. Many of those who have always owned will be young households who have only ever had one home.

3 The householder(s) is(are) the person(s) who bought/rented/found the house in the first place. Both men and women are counted as householders (i.e. households often have two householders). 'Female householder' (abbreviated in tables as 'woman') refers to either a woman living in a couple within a nuclear or extended household, or a woman who is a householder but has no resident partner.

4 A considerable number of respondents had income from sources other than employment, such as age or disability pensions or remittances from relatives. This income was often variable or irregular and so it was difficult to specify the amounts received. The proportion of households receiving such income varied between the groups studied. It was highest (22–35 per cent) amongst the older settlement owners; and lowest amongst young settlement owners or the tenants living in the older self-help settlement (6–10 per cent). The central-city tenants fell between the other groups (13–14 per cent). Income from such sources does not, therefore, seem likely to make a great difference to the comparison in Table 6.2. The figures for income 'in 1985 pesos' given in Table 6.2 were calculated by deflating the value at the time (month) of interview in accordance with the National Consumer Price Index for the month of interview compared with the overall 1985 index for the city in question.

5 Owners in El Salvador displayed the highest income heterogeneity of any group studied. This is because a number of households in the settlement are comparatively wealthy. They include, for example, one family who were cousins of the rector of the major Puebla university and who ran a small garment workshop on their plot. There was also a family who owned part of a quarrying operation taking place just beyond the settlement. This family had lived for a considerable period in the United States and clearly considered themselves different from other residents of El Salvador. The reason for the presence of such families is probably that El Salvador presents a number of attractions which are unusual for such a young irregular settlement. These result from an accident of its location (in a level area close to a major route into the city, in which wells can supply the residents' water requirements) and local politics (family ties between a settlement leader and an important member of the PRI in Puebla, and the existence of rival leaders who have sought to outdo each other in campaigning for services such as electricity and communal taxis).

6 The new values for Guadalajara were medians of 45,565 1985 pesos for young settlement owners, 40,780 pesos for older settlement tenants and 43,694 pesos for central city tenants.

7 A flat has been defined as an independent unit in a multi-occupied plot, i.e. one which does not require the household to enter common areas in order to make use of their toilet, bathroom or laundry facilities. By this definition, some housing has been identified as a 'flat' even though its general physical context would more readily identify it as part of a *vecindad*.

8 The very low possessions score for Buenos Aires is probably anomalous. The majority of possessions surveyed were electrical goods, and at the time three-quarters of homes in Buenos Aires lacked electricity.

9 The exact figures were: Agustín Yáñez, 6 per cent; Central Camionera, 5 per cent; Veinte de Noviembre, 16 per cent; Analco, 10 per cent.

10 If adult migrants are not excluded, differences in the balance of natives and migrants in the various groups studied will affect the averages.

11 For the purpose of this analysis, ten cases in which the household's present plot was not the first which it had owned were excluded. Also, these data refer only to households *purchasing* their plot, excluding those acquiring their plot by inheritance, or as a gift, etc. Figures are medians rounded up to the nearest whole number of years where appropriate.

12 These schemes, known as *tandas*, involve each member of the group paying a certain amount of money each month. Members then take it in turn to withdraw the entire collection for the month from the common fund.

13 The term 'single-parent families' is confined to nuclear households in Table 6.9, although some extended households may also be headed by a single parent. The reason for this is that in an extended household, even if the (male) spouse is absent, there may be other adults capable of helping with house construction. It should be noted, however, that if only the householder's immediate family within extended households is considered, 23 per cent of the extended households (5 per cent of all households surveyed) had a householder with resident children but no resident spouse (single-parent); and 26 per cent of extended households (5 per cent of all households surveyed) had a female householder with no resident spouse and with or without resident children (female-headed).

14 'Trader' excludes shop employees. It includes shop-owners, sales representatives and *vendedores ambulantes*. 'Construction worker' refers to a bricklayer or general labourer on a building site. 'Driver' includes taxi, bus and lorry drivers and chauffeurs. Certain industries (textiles, shoes and cars) were separated out from the category of 'factory worker' because of their regional importance in Jalisco or Puebla. 'Building trade crafts' refers to those practising specialised, construction-related, trades: carpenters, ironworkers, electricians, stonemasons, painters and plumbers. 'Police/security' includes nightwatchmen or security officers in private businesses (since the latter may be counted as part of the police force in Mexico).

15 A single average for *vecindad* tenants in the two settlements is given only for the sake of convenience. The figures should actually be given separately for each settlement: in Guadalajara, 4.4 for Agustín Yáñez *vecindad* dwellers, and 3.7 for those in Central Camionera; in Puebla, 4.4 and 5.2 respectively for Veinte de Noviembre and Analco.

16 It is also possible that the lack of room is one reason for *vecindad* tenants to register smaller resident households, since grown-up children may be more likely to seek accommodation elsewhere.

17 The two factors discussed are of course likely to be linked because in Mexico

rural dwellers have always been more likely to own their homes than their urban counterparts (see Chapter Three). It may be argued that it is incorrect not to distinguish between those who migrated to the city as children, and those who were adult migrants. However, excluding those who arrived in the city below the age of 16 years, the same results are obtained, for both men and women.

18  The couple described in the previous paragraph were from a small village near Chignahuapan in the northern *sierra* of the state of Puebla.

19  These points should not be interpreted to mean that rural migrants move directly to peripheral self-help ownership. As we have demonstrated, most had previously rented or shared accommodation in more consolidated parts of the city. The percentage of owner households interviewed consisting of couples who were already 'married' before they migrated to the city and who had moved directly to their present house varied from 2 per cent in Buenos Aires to 13 per cent in El Salvador; in both Guadalajara and Puebla, the figure for older settlement owners was 6 per cent. The comparatively high figure for El Salvador may be explained by a number of families who had moved from Chachapa (the village to which the *ejido* lands belonged), which was only just outside the city.

20  A small number of households have lived in the same house previously, but with a different kind of tenure. For example, one household previously lived as part of one of their parents' households, but now own the same property since their parents have moved away or died. Although such people clearly lived in the same area previously, they are excluded from Table 6.12, since there was a very specific reason for their remaining in the same area.

21  There is a marked tendency for people to move within the same sector of the city. This sectoral pattern of residential movements means that, if there is a scarcity of cheap land on the periphery of the city in the sector in which tenants are living, although cheap plots are plentiful on the other side of the city, tenants may simply not find out about opportunities open to them elsewhere in the city; as a result, they may fail to purchase a plot which they could afford. Perfect information about ownership opportunities cannot be assumed.

22  The permission to use the material collected by ITESO students was given to us by the Director of the School of Architecture; it is gratefully acknowledged.

23  The names given to the people described in case studies in Chapters Six and Seven are fictitious.

# 7  Landlords and the economics of landlordism

1  It should be noted, however, that the eight flats, which were large units with their own entrance from the street, occupied adjacent plots.

2  Evidence from a detailed survey of plot ownership in Analco based on records in the *Secretaría de Finanzas del Estado de Puebla*; all names are fictitious. The Alonso family – in reality, a number of interrelated families, descendants of several brothers owning property in the area in the 1940s – were the largest property-owners in Analco. Their holdings had varied over the years, reaching a maximum in the 1960s. Several members of the family still lived in the area; others owned their own homes elsewhere in the city. They also had at least one non-residential plot in Analco, used as a warehouse in 1986.

3 It is also possible that the property-owning Lebanese community in the city includes some powerful landlords. The Lebanese community tend to act as a closed group, in this as in other matters; they placed their rental property in the hands of the largest rental administrator in Puebla, who refused to provide information for this study.

4 When interviewed, an ex-governor of the state of Jalisco stated that in the 1950s landlords were normally wealthy, and owned a large number of properties. He quoted two individuals, one of whom, he said, had perhaps twenty *vecindades*. A representative of the Guadalajara tenants' union suggested that even in more recent years, over half the city's housing stock was in the hands of some 3,000 landlords, giving the improbably high figure of fifty properties per landlord.

5 It should be noted that Tables 7.1 and 7.3 cannot be interpreted as referring to the percentage of *landlords* living in particular places, etc., because more than one tenant could be interviewed in a particular property. This distorts the picture, exaggerating the characteristics of landlords with multi-tenant properties.

6 This accounts for the difference between the mean number of tenant households and the mean number of *all* households living on the rental properties surveyed in the case-study settlements (Table 7.4); a difference which is particularly noteworthy in the case of Puebla.

7 See Case Study 5 in the Appendix to this chapter.

8 Twenty-four landlords of the thirty-seven landlords in Puebla reported receiving less than the minimum salary from the rent.

9 See for example Case Study 5 in the Appendix to this chapter.

10 Inheritance may also account for the large numbers of single-tenant properties, particularly in Guadalajara, and particularly in the older self-help areas. As many as 11 per cent of (all) owners surveyed in Agustín Yáñez, and 25 per cent of those in Veinte de Noviembre, had acquired their property through inheritance or as a gift, usually from their parents. These were resident owners; but the same may well apply to owners who let their inherited homes to a single tenant household. They may keep the property for sentimental reasons, or for later use by their own children. The rent may also serve to supplement their income. Comments made by the tenants of such properties confirm that this is a reasonable interpretation.

11 For many centuries, Puebla grew only slowly and the eastern edge of the *barrio* was effectively the edge of the city. During the 1940s, La Carbonera, an agricultural property on the edge of the *barrio*, was subdivided and sold in blocks which now constitute the easternmost blocks of the area designated as Analco. The 'self-help landlords' are mostly located in this area.

12 Méndez Rodríguez (1987) carried out a study of the rents advertised in two major Mexico City newspapers. He observes that whereas the average rent in 1976 required 27 days of earnings at the minimum salary, by 1986 it had risen to the equivalent of 67 days earnings (ibid.: 94). We have two major doubts about the validity of this finding. First, the results do not control for changes in the location or quality of the accommodation. It is possible that both changed considerably between 1976 and 1986, thereby distorting his average rent figure. Second, there are reasons to doubt that the rents advertised in the newspapers are the rents finally negotiated: indeed, we wonder who is

advertising property for rent when so much low-income accommodation is let by word of mouth.

13  This was the rental income of a man with sixteen good-quality flats in Analco.

14  Some landlords in Puebla followed what used to be a legal requirement to register their rental contracts with the state Treasury; several of them said that they had been told they did not need to pay tax because the income they received from renting was too low.

15  See Case Study 2 in the Appendix to this chapter.

16  A comment made by a North American ex-patriot who had settled in Guadalajara. He runs a small estate agency, aimed particularly at United States citizens who come to the city to retire. He felt that 'there is no economic justification for being in the renting business'.

17  He was also lending money to another employee to buy his house, for the same reason.

18  Bank loans for the purchase of existing property were reported to have been suspended at the time of this study.

19  The President of the *Asociación Mexicana de Profesionales Inmobiliarios*, a Puebla-based association of rental administrators and property developers, claimed that in the past it had been difficult to get people to accept the idea of co-ownership involved in a condominium, but that economic circumstances were now obliging them to do so. Tenants were at first unwilling to buy their rented home, because they did not like to take the risks associated with ownership. Eventually, however, half of those buying their homes are likely to be existing tenants.

20  The frequency of commercial transactions affecting property in Analco is in fact over-estimated by these figures, since only about one-half of the changes of ownership involve entirely commercial transactions (the others involve transfers of property between different members of the same family, by inheritance, 'sale', or gift, for example). In all, we may suggest that perhaps one-fifth of the properties in Analco were exchanged on the open market in the last fifteen years; one-third, in the last twenty-five years.

21  The irresponsibility of certain owners in this respect was indicated by a number of anecdotes told at a meeting of rental administrators and property agents in Puebla; they concluded that there were some owners who would rather see their property destroyed, even against their own economic interests, rather than submit to government regulations concerning preservation of historic monuments.

22  Some detailed case histories are given in Universidad Autónoma de Puebla, DIAU-ICUAP (1984).

23  Presentation by Raúl Contreras, of the Universidad Autónoma de Puebla, to a workshop on 'Las ciudades mexicanas: historia y sociedad', Universidad Veracruzana, Xalapa, April 1986.

24  Presentation by Dr Efraín Castro Morales, former head of the *Instituto Nacional de Antropología e Historia* in Puebla, to meeting of the *Asociación Mexicana de Profesionales Inmobiliarias*, August 1986.

25  The difference was smaller in Agustín Yañez, where plots with owners and tenants accounted for 3 per cent of plots; those with owners and sharers, 4 per cent.

# 8 Landlord–tenant relations

1 Newly-formed households seem to stay in their first home for a relatively short period before moving on to better quality, probably more expensive, rented accommodation. This is a parallel to the short period of time spent by young couples living as members of a parental household before finding independent accommodation; some households go through both these stages before finding more permanent rented accommodation. Nevertheless, even for earlier homes the average length of tenancy is quite long.

2 Mean value quoted; the median was 8 years, and individual tenancies ranged from 2 to 30 years.

3 Mean value quoted; the median was 19 years; the range was 4 to 42 years.

4 Comparing the length of tenants' residential histories with the time spent in their current house reveals a high and significant direct correlation between them. The importance of this relationship may be judged from Table 8.3, showing the notable consistency with which the time in the present house increases with the length of time since the household was formed.

5 The difference between the two cities appears to be due to the greater frequency with which tenants were evicted in Guadalajara in order for the house to be sold or demolished (or, less frequently, repaired). This possibly reflects the more frequent changes of use in the central area of Guadalajara compared with its Puebla counterpart. Conversely, tenants in Puebla more frequently left of their own accord in order to move to a better or larger house. This may reflect the poorer housing conditions in the central area of Puebla, particularly in the *vecindades*.

6 In both cities, it is legal to pay the rent at the end of the month, rather than the beginning.

7 This may be why a higher proportion of tenants in Guadalajara reported that they had been evicted for this reason of major repairs. Although the proposed changes did not always take place in practice, it is, in theory, risky for Jalisco landlords to *pretend* that they are going to occupy or repair the property. If they do not in fact do so within six months, they may legally be required to pay damages (equivalent to at least six months' rent) to the tenant. There does not seem to be any equivalent regulation in Puebla.

8 After that time, they may still, with the landlord's permission, remain legally as tenants: in Jalisco, they are then considered to be renting for an 'unlimited period' (*tiempo indefinido*), and the landlord has to give one year's notification if he or she wishes the tenant to leave. In Puebla, it then appears to be up to the tenant to decide when to leave, although the *Código Civil* is not very clear on this point. In both states, tenants must continue to pay the rent.

9 Interview with representative of *Federación de Cámaras de la Propiedad Urbana de Jalisco*, January 1986.

10 A representative of the *Federación de Cámaras de la Propiedad Urbana de Jalisco* reported that some tenants would either invent the name of a witness or give the name of a witness who lived out of the country. This would delay settlement of the case.

11 As many as 14 per cent of Central Camionera tenants were doing so, although these were mostly tenants in a single large *vecindad* whose tenants were in conflict with the owners. Tenants cannot deposit the rent with the courts at

will: there has to be some reason why they cannot pay the rent directly to the owner, as usual.

12 Some court cases nevertheless continue, after such an arrangement, because the tenant does not in fact move on the date agreed. It is in these circumstances that legal officers are most likely to be sent in to evict the tenant and seize goods to make good some of the landlord's loss.

13 This was confirmed by the rental administrators in Puebla, who had different opinions of the practice. Some argued that it was merely a matter of facing up to the realities of the situation; others, that it was a bad practice, because it affected the market adversely and is a form of corruption. Even critics of the practice, however, clearly used it when it suited them.

14 This was reported by a left-wing lawyer in Puebla who appeared for tenants in their cases with landlords.

15 The names of landlords quoted in this chapter refer back to the landlord case studies in Chapter Seven.

16 The remaining 3 per cent had known about the property through other means. The data from all the settlements were very similar, so they have not been disaggregated.

17 There is a certain discrepancy between landlord and tenant accounts in this matter. Many more tenants reported a previous acquaintance with the landlord, whereas landlords reported that tenants were mostly unknown to them. This may perhaps be explained as a result of tenants with a mutual acquaintance with the landlord being counted as 'unknown' by the landlord. The differences between the figures on how tenants came to know that a house was to let and Table 8.5 are a product of different sample sizes.

18 In the Analco *vecindad* whose tenants were studied in detail, two of the households were related, as the married son of one resident also lived there with his family; in the past, moreover, there had been three generations residing there, since the son's grandmother had also been a tenant for many years. Marroquín (1985: 209) also found different households belonging to extended families in the larger *vecindades* in central Puebla.

19 In Guadalajara, around three-quarters of the tenants interviewed reported that the landlord gave them receipts for their rent payments; in Puebla, well in excess of four-fifths of the tenants interviewed were given receipts.

20 This is a practice explicitly recognised in the *Código Civil* of each state.

21 The handful of families living in a sublet property or a *casa de huespedes* had rents quoted in weekly, or even daily, terms. There were no contracts in these or other cases of subletting. The maximum period for which contracts can be issued is fifteen years in Jalisco, and ten years in Puebla; but when the contract ends, tenants who are not in arrears have the right to a *prórroga*.

22 The exact figures were 41 per cent in Agustín Yáñez, 40 per cent in Central Camionera, 28 per cent in Veinte de Noviembre, and 16 per cent in Analco. However, 12 per cent of Central Camionera tenants reported that they had previously had a contract with the landlord, and the existence of an earlier contract gives tenants certain rights; as a result, these 12 per cent may actually have been in the same position as tenants in the other settlements who reported that they did have a contract.

23 The landlord is also required to pay damages for any injury or other nuisance caused by defects in a property which existed before the tenant occupied the building.

24 For example, the state Treasurer had intended to increase the property tax rate to over twice its present level. The landlords' organisation contested this and eventually won a *reduction* in the rate.

25 The AMPI was founded in the 1950s by rental administrators in Puebla, and the city still has one of the largest number of administrators who are members of the Association.

26 Its leader, a department head in the municipal authorities, was also unable to name specific cases or areas in which they had been active. He stressed that the organisation does *not* seek confrontation with landlords. The man in charge of the Federation's office in Central Camionera argued that the landlords were the party most aggrieved by the existing rental legislation.

27 Information provided by one of the original leaders of this movement. Castillo (1986: 327–33) states that the leaders were actually unable to achieve their minimum target of forty families from the tenants' movement to act as a core for the invasion. As in other cases, the arrival of 'non-politicised' families also wishing to gain a plot in the invasion distorted the action from its original political intentions.

28 This action was taken because plots in the scheme, with government finance, were being allocated to the PRI tenants' organisation.

29 One leader reported this happening in the case of the Xonaca invasion. Castillo (1986: 300) uses the case of the 1972 foundation of *colonia* Pablo Juárez Ruiz by another Puebla tenants' organisation to show how its very success in getting land for members had undermined the association as the new home-owners lost interest in its aims. For Guadalajara, Durand (1984: 16) also concludes that tenants' organisations are weakened by their members' concentration on short-term goals.

30 The owner had recently died, leaving the property to her children, who lived in Mexico City and were not interested in renting; they wanted to evict all the tenants and sell the property. Twelve of the fifty families in the *vecindad* had signed a new contract giving them a few more months to live there, after which they had formally agreed to leave; but the owners had taken legal action against nineteen other families in an attempt to scare the rest into leaving. The residents had then sought the help of the UII, although not all of the tenants agreed that it should be involved. They illustrate the complaint made by tenants' organisations that tenants only approach the association when they have an immediate problem (usually an eviction threat) facing them.

31 Little has been written about the tenants' movement in Puebla prior to the 1940s. However, the potential of housing issues to fuel political unrest is demonstrated by the inclusion of lower rents and better housing conditions among Puebla textile workers' demands during a period of strikes in the early years of the Revolution (LaFrance, 1983).

32 Other commentators do not necessarily accept the importance of this movement in terms of the participation of tenants who were not previously politically active. The movement of 1922 did achieve considerable popular support in Veracruz and Mexico City, however (García Mundo, 1976; Taibo and Vizcaíno, 1984).

33 A representative of the Federation of Chambers of Urban Property confirmed that tenants' organisations posed little threat to their members.

## 9 The future of renting: policy options

1 The term 'political rationality' has been described, rather than defined, by Diesing (1962) as follows:

> In a political decision ... action never is based on the merits of a proposal but always on who opposes it ... A course of action which corrects economic or social deficiencies but increases political difficulties must be rejected, while an action which contributes to political improvement is desirable even if it is not entirely sound from an economic and social standpoint.

See Gilbert and Ward (1985: 149–53) for further discussion.
2 Discussions about landlords with René Coulomb.
3 In Sri Lanka there is a national repairs fund (UN, 1979: 37).
4 Coulomb recommends the establishment of a *Fondo para el Mejoramiento de Vivienda en Arrendamiento del D.F.*

# Bibliography

Aaron, H. (1966) 'Rent controls and urban development: a case study of Mexico City', *Social and Economic Studies*, 15, 314–28.

Abrams, C. (1964) *Man's Struggle for Shelter in an Urbanizing world*, London: The MIT Press.

Abt Associates Inc. (1982) 'Informal housing in Egypt', mimeo.

Aguirre, C. and Carabarín, A. (1983) 'Propietarios de la industria textil de Puebla en el siglo XIX: Dionisio José de Velasco y Pedro Berges de Zúñiga', in CIHS (Centro de Investigaciones Históricas y Sociales), 177–224.

Alba, C. (1986) 'La industrialización en Jalisco: evolución y perspectivas', in G. de la Peña and A. Escobar (eds), 89–146.

Alonso Palacios, A. (1983) *Los Libaneses y la Industria Textil en Puebla*, Puebla: Centro de Investigaciones y Estudios Superiores en Antropología Social, Cuadernos de la Casa Chata.

Amis, P. (1982) 'Squatters or tenants?: the commercialisation of unauthorised housing in Nairobi', paper presented to the Annual Conference of the Development Studies Association, Dublin.

Amis, P. (1987) 'Migration, urban poverty, and the housing market: the Nairobi case', in J. Eades (ed.), *Migrants, Workers, and the Social Order*, London: Tavistock publications, 249–68.

Angel, S., Archer, R.W., Tanphiphat, S. and Wegelin, E.A. (eds) (1983) *Land for Housing the Poor*, Singapore: Select Books.

Arias, P. (1985) 'La industria en perspectiva', in P. Arias (ed.), 77–130.

Arias, P. (ed.) (1985) *Guadalajara: la Gran Ciudad de la Pequeña Industria*, El Colegio de Michoacán.

Arias, P. and Roberts, B. (1985) 'The city in permanent transition: the consequences of a national system of industrial specialization', in J. Walton (ed.) *Capital and Labor in the Industrialized World*, Newbury Park, CA.: Sage, 149–75.

Armus, D. and Hardoy, J. (1984) 'Vivienda popular y crecimiento urbano en el Rosario del novecientos', *Revista Latinoamericana de Estudios Urbano Regionales (EURE)*, 31, 29–54.

Arroyo, J. (1985) 'Ires y venires en el Occidente', in P. Arias (ed.), 21–56.

Ayuntamiento de Guadalajara (1955) *Evolución de Guadalajara*, Guadalajara.

Azuela, A. (1987) 'De inquilinos a propietarios. Derecho y política en el Programa de Renovación Habitacional Popular', *Estudios Demográficos y Urbanos*, 2, 53–74.

Azuela, A. (1989) *La Ciudad, la Propiedad Privada y el Derecho*, El Colegio de México.

Bähr, J. and Mertins, G. (1985) 'Desarrollo poblacional en el Gran Santiago entre 1970 y 1982: análisis de resultados censales en base a distritos', *Revista de Geografía Norte Grande*, 12, 11–26.

Balán, J., Browning, H.L. and Jelin, E. (1973) *Men in a Developing Society: Geographic and Social Mobility in Monterrey, Mexico*, University of Texas Press.

Barnes, S.T. (1982) 'Public and private housing in urban West Africa: the rental implications', in W.K.C. Morrison and P.C.W. Gutkind (eds) *Housing the Urban Poor in Africa*, Syracuse University, 5–32.

Baross, P. (1983) 'The articulation of land supply for popular settlements in Third World cities', in S. Angel, R.W. Archer, S. Tanphiphat, and E.A. Wegelin (eds), 180–209.

Baross, P. (1987) 'Land supply for low-income housing: issues and approaches', *Regional Development Dialogue*, 8, 29–45.

Bazant, J. (1977) *Los Bienes de la Iglesia en México 1856–1875: Aspectos Económicos y Sociales de la Revolución Liberal*, El Colegio de México.

Beijaard, F. (1986) *On Conventillos: Rental Housing in the Centre of La Paz, Bolivia*, Amsterdam: Free University of Amsterdam, Urban Research Working Paper No. 5.

Bennett, V. (1989) 'Urban public services and social conflict: water in Monterrey', in A.G. Gilbert (ed.) *Housing and Land in Urban Mexico*, University of California, San Diego, Center for US–Mexican Studies, 79–100.

Benton, L.A. (1987) 'Reshaping the urban core: the politics of housing in authoritarian Uruguay', *Latin American Research Review*, 22, 33–52.

Berthe, J.-P. (1973) 'Introducción a la historia de Guadalajara y su región', in Institut des Hautes Etudes de l'Amérique Latine, 130–47.

Blaesser, B.W. (1981) *Clandestine Development in Colombia*, USAID Office of Housing, Occasional Paper Series.

Boleat, M. (1985) *National Housing Finance Systems: a Comparative Analysis*, London: Croom Helm.

Bortz, J. (1984) 'Industrial wages in Mexico City 1939–75', unpublished PhD thesis, University of California, Los Angeles.

Brennan, E.M. (1978) 'Demographic and social patterns in urban Mexico: Guadalajara', 1876–1910, unpublished PhD thesis, Columbia University.

Brown, J. (1972) *Patterns of Intra-urban Settlement in Mexico City: an Examination of the Turner Theory*, New York: Cornell University Latin American Studies Program, Dissertation Series 40.

Burgess, R. (1982) 'Self-help housing advocacy: a curious form of radicalism: a critique of the work of John F.C. Turner', in P.M. Ward (ed.), 56–97.

Butterworth, D. and Chance, J. (1981) *Latin American Urbanization*, Cambridge University Press.

Carroll, A. (1980) *Private Subdivisions and the Market for Residential Lots in Bogotá*, City Study, Washington: World Bank.

Castells, M. (1977) *The Urban Question*, London: Edward Arnold.

Castells, M. (1983) *The City and the Grassroots*, London: Edward Arnold.

Castillo, J. (1986) 'El movimiento urbano popular en Puebla', in M. García, M. and J. Castillo (eds) (1986) *Los Movimientos Sociales en Puebla* Tomo 2, Universidad Autónoma de Puebla, 201–360.

CENVI (Centro de la Vivienda e Estudios Urbanos) (1986) *Instituciones Públicas y Organizaciones Sociales Frente al Mejoramiento de Cuatro Asentamientos en la Ciudad de México*, Medellín, Colombia: Centro de Estudios del Habitat Popular (CEHAP).

CENVI (Centro de la vivienda e estudios urbanos) (1989) 'Vivienda y propiedad en cinco colonias populares de la ciudad de México, Primer documento de interpretación', mimeo.

Cerutti, M. (ed.) (1985) *El Siglo XIX en México – Cinco Procesos Regionales: Morelos, Monterrey, Yucatán, Jalisco y Puebla*, Mexico City: Claves Latinoamericanos.

CEU (Centro de Estudios Urbanos) (1989) 'Papel de trabajo: primeros avances de la investigación "El inquilinato y la vivienda compartida de ciudades latinoamericanas" ', mimeo.

CIHS (Centro de Investigaciones Históricas y Sociales) (1983) *Puebla en el Siglo XIX: Contribución al Estudio de su Historia*, Universidad Autónoma de Puebla.

Clapham, D., Kintrea, K. and Munro, M. (1987) 'Tenure choice: an empirical investigation', *Area*, 15, 11–18.

Clark, W.A.V. and Onaka, J.L. (1983) 'Life cycle and housing adjustment as explanations of residential mobility', *Urban Studies*, 20, 47–57.

Cleaves, P. (1974) *Bureaucratic Politics and Administration in Chile*, University of California Press.

Collier, D. (1976) *Squatters and Oligarchs*, London: Johns Hopkins University Press.

Colombia, DANE (Departamento Administrativo Nacional de Estadística), (1977) *La Vivienda en Colombia*, Bogotá.

Connolly, P. (1982) 'Uncontrolled settlements and self-build: what kind of solution? The Mexico City case', in P.M. Ward (ed.), 141–74.

Connolly, P. (1987) 'La política habitacional después de los sismos', *Estudios Demográficos y Urbanos*, 2, 101–20.

Consumers' Association (1985) *Renting and Letting*, London: Consumers' Association.

Contreras, C. and Gross, J.C. (1983) 'La estructura ocupacional y productiva de la ciudad de Puebla en la primera mitad del siglo XIX', in CIHS (Centro de Investigaciones Históricas y Sociales), 111–76.

Conway, D. and Brown, J. (1980) 'Intraurban relocation and structure: low-income migrants in Latin America and the Caribbean', *Latin American Research Review*, 15, 95–125.

COPEVI (Centro Operacional de Vivienda y Poblamiento) (1977) *Investigación Sobre la Vivienda II: la Producción de Vivienda en la Zona Metropolitana de la Ciudad de México*, Mexico City.

Cornelius, W. (1975) *Politics and the Migrant Poor in Mexico City*, Stanford University Press.

Cornelius, W., Gentleman, J. and Smith, P.H. (eds) (1989) *Mexico's Alternative Political Futures*, Center for US–Mexican Studies, University of California, San Diego, Monograph Series, 30.

Coulomb, R. (1981) 'La producción y consumo de la vivienda en alquiler con especial referencia al Area Metropolitana de la Ciudad de México', mimeo.

Coulomb, R. (1985a) *La Legislación en Materia de Vivienda en Arrendamiento para el Distrito Federal: Situación Actual (1985) y Propuestas Reglamentarias*, Cuadernos del CENVI, Mexico City.

220    Bibliography

Coulomb, R. (1985b) 'La vivienda de alquiler en las áreas de reciente urbanización', A: *Revista de Ciencias Sociales y Humanidades*, VI, 43–70.

Coulomb, R. (1989) 'Rental housing and the dynamics of urban growth in Mexico City', in A.G. Gilbert (ed.), 39–50.

Cruz Rodríguez, M.S. (1982) 'El ejido en la urbanización de la Cd. de México', *Habitación: Problemas de Vivienda y Urbanismo*, 2, 29–43.

Danielson, M.N. and Keles, R. (1984) *The Politics of Rapid Urbanization in Turkey*, New York: Holmes and Meier.

Daunton, M.J. (1987) *A Property Owning Democracy? Housing in Britain*, London: Faber and Faber.

Diesing, P. (1962) *Reason in Society*, University of Illinois Press.

Dietz, H.A. (1980) *Poverty and Problem Solving under Military Rule: the Urban Poor in Lima, Peru*, University of Texas Press.

Doebele, W. (1987) 'The evolution of concepts of urban land tenure in developing countries', *Habitat International*, 11, 7–22.

Doling, J. (1976) 'The family life cycle and housing choice', *Urban Studies*, 13, 55–8.

Dotson, F. and Dotson, L.O. (1953) 'Ecological trends in the city of Guadalajara, Mexico', *Social Forces*, 31, 367–74.

Downs, A. (1983) *Rental Housing in the 1980s*, Washington: The Brookings Institution.

Drakakis-Smith, D.W. (1976) 'Slums and squatters in Ankara: case studies in four areas of the city', *Town Planning Review*, 47, 225–40.

Dreier, P. (1984) 'The tenants' movement in the United States', *International Journal of Urban and Regional Research*, 8, 255–79.

Duhau, E. (1987) 'La formación de una política social; el caso del Programa de Renovación Habitaciónal Popular', *Estudios Demográficos y Urbanos*, 2, 75–100.

Duhau, E. (1988) 'Política habitacional para los sectores populares en México. La experiencia de FONHAPO', *Medio Ambiente y Urbanización*, 7, 34–45.

Durand, J. (1984) 'El movimiento inquilinario de Guadalajara, 1922', *Encuentro*, 1(2), 7–28.

Durand-Lasserve, A. (1986) *L'Exclusion des Pauvres dans les Villes du Tiers-monde*, Paris: L'Harmattan.

Edwards, M.A. (1981) 'Cities of tenants: renting as a housing alternative among the Colombian urban poor', unpublished PhD thesis, University College London.

Edwards, M.A. (1982) 'Cities of tenants: renting among the urban poor in Latin America', in A.G. Gilbert, J.E. Hardoy and R. Ramírez (eds), 129–58.

El Kadi, G. (1988) 'Market mechanisms and spontaneous urbanization in Egypt: the Cairo Case', *International Journal of Urban and Regional Research*, 12, 22–37.

Englander, D. (1983) *Landlord and Tenant in Urban Britain, 1834–1918*, London: Clarendon Press.

Escobar, A. (1986a) *Con el Sudor de tu Frente: Mercado de Trabajo y Clase Obrera en Guadalajara*, El Colegio de Jalisco.

Escobar, A. (1986b) 'Patrones de organización social en el mercado de trabajo manual de Guadalajara', in G. de la Peña and A. Escobar (eds), 147–89.

Escobar, A. (1988) 'The rise and fall of an urban labour market: economic crisis and the fate of small workshops in Guadalajara, Mexico', *Bulletin of Latin American Research*, 7, 183–205.

Estrada Urroz, R. (1986) 'El problema de la vivienda y las luchas inquilinarias en Puebla, 1940–60', in Universidad Autónoma de Puebla (DIAU–ICUAP), 143–58.

Foster, J. (1979) 'Review of Wohl, A., *How Imperial London Preserved its Slums'*, *International Journal of Urban and Regional Research*, 3, 93–114.

Gabayet, L, Lailson, S. and Padilla, C. (1987) 'El sector informal en Guadalajara: tres estudios de caso', report prepared for the International Labour Office (ILO), mimeo.

Gamboa, L. (1985) *Los Empresarios de Ayer: el Grupo Dominante en la Industria Textil de Puebla 1906–1929*, Universidad Autónoma de Puebla.

García Barragán, M. (1947) *Memoria del Poder Ejecutivo 1943–47*, Guadalajara: Artes Gráficas.

García Mundo, O. (1976) *El Movimiento Inquilinario de Veracruz, 1922*, Mexico City: SepSetentas.

García Peralta, B. (1986) 'La lógica de las grandes acciones inmobiliarias en la ciudad de Querétaro', *Estudios Demográficos y Urbanos*, 1, 375–98.

Garza, G. and Schteingart, M. (1978) *La Acción Habitacional del Estado en México*, El Colegio de México.

Geisse, G. and Sabatini, F. (1982) 'Urban land market studies in Latin America: issues and methodology', in M. Cullen and S. Woolery (eds) *World Congress on Land Policy*, London: Lexington Books, 147–76.

Gertz Manero, F. (1964) *La Vivienda Congelada en el Distrito Federal*, Mexico: Porrúa.

Gilbert, A.G. (1981) 'Pirates and invaders: land acquisition in urban Colombia and Venezuela', *World Development*, 9, 657–78.

Gilbert, A.G. (1983) 'The tenants of self-help housing: choice and constraint in the housing markets of less developed countries', *Development and Change*, 14, 449–77.

Gilbert, A.G. (1987) 'Urban unrest in Latin America', in A. Gauhar (ed.) *Third World Affairs 1987*, London: Third World Foundation for Social and Economic Studies, 405–13.

Gilbert, A.G. (1989a) 'Housing during recession: illustrations from Latin America', *Housing Studies*, 4, 155–66.

Gilbert, A.G. (1989b) 'Rental housing in developing countries', report produced for UNCHS (Habitat), mimeo.

Gilbert, A.G. (ed.) (1989c) *Housing and Land in Urban Mexico*, University of California, San Diego, Center for US–Mexican Studies.

Gilbert, A.G. and Gugler, J. (1982) *Cities, Poverty and Development: Urbanization in the Third World*, Oxford: Oxford University Press.

Gilbert, A.G., Hardoy, J.E. and Ramírez, R. (eds) (1982) *Urbanization in Contemporary Latin America*, Chichester: John Wiley and Sons Ltd.

Gilbert, A.G and Healey, P. (1985) *The Political Economy of Land: Urban Development in an Oil Economy*, Aldershot: Gower.

Gilbert, A.G. and van der Linden, J. (1987) 'The limits of a Marxist theoretical framework for explaining state self-help housing', *Development and Change*, 18, 129–36.

Gilbert, A.G. and Varley, A. (1988) 'Housing tenure and the urban poor in Third World cities', final report to Overseas Development Administration, mimeo.

Gilbert, A.G. and Ward, P.M. (1982) 'Residential movement among the urban poor: the constraints on housing choice in Latin American cities', *Transactions, Institute of British Geographers*, 7, 129–49.

Gilbert, A.G. and Ward, P.M. (1985) *Housing, the State and the Poor: Policy and Practice in Three Latin American Cities*, Cambridge: Cambridge University Press.

González, V. (1980) 'La industria textil en Puebla, 1960–76', *Boletín de Investigación del Movimiento Obrero*, 1, 72–88.

González Navarro, M. (1974) *Población y Sociedad en México*, Universidad Nacional Autónoma de México.

González de la Rocha, M. (1984) 'Domestic organization and the reproduction of low income households: the case of Guadalajara, Mexico', unpublished PhD thesis, University of Manchester.

González de la Rocha, M. (1986a) 'Lo público y lo privado: el grupo doméstico frente al mercado de trabajo urbano', in G. de la Peña and A. Escobar (eds), 191–233.

González de la Rocha, M. (1986b) *Los Recursos de la Pobreza: Familias de Bajos Ingresos de Guadalajara*, El Colegio de Jalisco.

Gormsen, E. (1978) 'La zonificación socio-económica de la ciudad de Puebla: cambios por efecto de la metropolización', *Comunicaciones*, 15, 7–22.

Green, G.S. (1988a) 'Finding a home in a frontier city: the dynamics of housing tenure in Santa Cruz, Bolivia', unpublished PhD thesis, University College London.

Green, G.S. (1988b) 'The quest for *tranquilidad*: paths to home ownership in Santa Cruz, Bolivia', *Bulletin for Latin American Research*, 7 (1), 1–16.

Grennel, P. (1972) 'Planning for invisible people: some consequences of bureaucratic values and practices', in J.F.C. Turner and R. Fichter (eds) *Freedom to Build*, Don Mills, Ontario: Collier Macmillan, 95–121.

Grimes, O.F. (1976) *Housing for Low-Income Urban Families: Economics and Policy in the Developing World*, London: Johns Hopkins University Press.

Grosso, J.C. (1984) *Estructura Productiva y Fuerza de Trabajo, Puebla, 1830–1890*, Puebla: Cuadernos de la Casa Presno.

Grosso, J.C. (1985) 'Estructura productiva y fuerza de trabajo en el área del municipio de Puebla (siglo XIX)', in M. Cerutti (ed.), 200–39.

Hamer, A.M. (1981) *Las Subdivisiones no Reglamentadas de Bogotá: los Mitos y Realidades de la Construcción de Vivienda Suplementaria*, Bogotá: Corporación Centro Regional de Población, La Ciudad Documentada No. 24.

Hamnett, C. (1986) 'Who gets to own?', *New Society*, 25 July, 18–19.

Handelman, H. (1975) 'The political mobilization of urban squatter settlements', *Latin American Research Review*, 10, 35–72.

Harloe, M. (1985) *Private Rented Housing in the U.S. and Europe*, London: Croom Helm.

Hernández, E. and Parás, M. (1988) 'México en la primera década del siglo XXI: las necesidades sociales futuras', *Comercio Exterior*, 38, 963–78.

Hernández Laos, E. (1984) 'La desigualdad regional en México (1900–1980)', in R. Cordera and C. Tello (eds) *La Desigualdad en México*, Mexico City: Siglo Veintiuno Editores, 155–92.

Herzog, L. (1989) 'Tijuana: state intervention and urban form in a Mexican border city', in A.G. Gilbert (ed.), 109–33.

Hoenderdos, W. (ed.) (1985) *Migración, Empleo y Vivienda en la Ciudad de Chihuahua*, Chihuahua: Desarrollo Económico del Estado de Chihuahua.

Hoenderdos, W., van Lindert, P. and Verkoren, O. (1983) 'Residential mobility, occupational changes and self-help housing in Latin American cities: first impressions from a current research programme', *Tijdschrift voor Economische et Sociale Geografie*, 74, 376–86.

Howenstine, E.J. (1983) 'Rental housing in industrialized countries: issues and policies', in J.C. Weicher *et al.* (eds) *Rental Housing: Is There a Crisis?*, Washington: The Urban Institute Press, 99–110.

IADB (Interamerican Development Bank) (1988) *Economic and Social Progress in Latin America*, Washington: Interamerican Development Bank.

Ibañez, E. and Vázquez, D. (1970) *Guadalajara: un Análisis Urbano*, Ediciones de la Comisión de la Coordinación Urbana del Valle de Atemajac.

Institut des Hautes Etudes de l'Amerique Latine (1973) *Regiones y Ciudades en América Latina*, Mexico City: SepSetentas.

ISS (Institute of Social Studies) (1989) Workshop on Rental Housing in Indonesia, 23–27 October, The Hague.

ITESO (Instituto Tecnológico de Estudios Superiores del Occidente) (1984) *Alternativas de vivienda en el Area Central*, Programa de Estudios Urbanos.

Jalisco, DPUEJ (Departamento de Planeación Urbano del Estado de Jalisco) (1979) *Estudio de Funciones Urbanas en el Area Metropolitana de Guadalajara*, Gobierno de Jalisco.

Jalisco, DPUEJ/SAHOP (Secretaría de Asentamientos Humanos e Obras Públicas) (1981) *Plan de Ordenamiento Urbano de la Zona Conurbada de Guadalajara*.

Jalisco, DPUEJ/SEDUE (Secretaría de Desarrollo Urbano y Ecología) (1985) *Plan de Ordenamiento de la Zona Conurbada de Guadalajara: Planes Parciales de Urbanización y Control de la Edificación*.

Jones, G. (1989) 'The commercialization of the land market? Land ownership patterns in the city of Puebla, Mexico', mimeo.

Karst, K., Schwartz, M. and Schwartz, A. (1973) *The Evolution of the Law in the Barrios of Caracas*, University of California Press.

Keare, D.A. and Parris, S. (1982) *Evaluation of Shelter Programs for the Urban Poor: Principal Findings*, Washington: World Bank Staff Working Papers No. 547.

Keles, R. and Kano, H. (1987) *Housing and the Urban Poor in the Middle East – Turkey, Egypt, Morocco and Jordan*, Tokyo: Institute of Developing Economies, MES Series No. 20.

Kemp, P. (1982) 'Housing landlordism in late nineteenth century Britain', *Environment and Planning A*, 14, 1437–47.

Kemp, P. (1987) 'Some aspects of housing consumption in late nineteenth century England and Wales', *Housing Studies*, 2, 3–16.

Kemper, R.V. and Royce, A.P. (1981) 'La urbanización mexicana desde 1821: un enfoque macro-histórico', *Relaciones: Estudios de Historia y Sociedad*, 2, 5–39.

Kendig, H.L. (1984) 'Housing careers, life cycles and residential mobility: implications for the housing market', *Urban Studies*, 21(3), 271–83.

Kowarick, L. (ed.) (1988) *As Lutas Sociais e a Cidade: São Paulo: Passado e Presente*, São Paulo: Paz e Terra.

Kruijt, D. and Alba, C. (1988) 'Grupos empresariales y poder regional en el occidente de México: la burguesía tapatia', paper presented to 46th International Congress of Americanists, Amsterdam.

LaFrance, D.G. (1983) 'Los obreros y la Revolución Mexicana: el Presidente Francisco I. Madero y los Trabajadores Textiles de Puebla', *Boletín de Investigación del Movimiento Obrero*, 3, 6, 29–49.

Laun, J.I. (1977) 'El estado y la vivienda en Colombia: análisis de urbanizaciones del Instituto de Crédito Territorial en Bogotá', in C. Castillo (ed.) *Vida Urbana y Urbanismo*, Instituto Colombiano de Cultura, Biblioteca Básica de Colombia, vol. 30.

Legorreta, J. (1983) *El Proceso de Urbanización en Ciudades Petroleras*, Mexico City: Centro de Ecodesarrollo.

Lemer, A.C. (1987) *The Role of Rental Housing in Developing Countries: a Need for Balance*, Washington: World Bank, Water Supply and Urban Development Department, Discussion Paper, Report No. UDD–104.

Liehr, R. (1976) *Ayuntamiento y Oligarquía en Puebla, 1787–1810*, Mexico City: SepSetentas.

van der Linden, J. (1987) *The Sites and Services Approach Reviewed*, Aldershot: Gower.

Lindley, R.B. (1983) *Haciendas and Economic Development: Guadalajara, Mexico, at Independence*, University of Texas Press.

Logan, K. (1979) 'Migration, housing, and integration: the urban context of Guadalajara, Mexico', *Urban Anthropology*, 8, 2, 131–48.

Logan, K. (1984) *Haciendo Pueblo: the Development of a Guadalajara Suburb*, University of Alabama Press.

Loreto, R. (1986) 'La distribución de la propiedad en la ciudad de Puebla en la década de 1830', in Universidad Autónoma de Puebla (DIAU–ICUAP), 27–41.

Losada, R. and Gómez, H. (1976) *La Tierra en el Mercado Pirata de Bogotá*, Bogotá: FEDESARROLLO.

Lugo, M. (1989) 'Vivienda en arrendamiento: problemática y política gubernamental', México, SEDUE, mimeo.

McAuslan, P. (1985) *Urban Land and Shelter for the Poor*, London: Earthscan, International Institute for Environment and Development.

MacLennan, D. (1982) *Housing Economics: an Applied Approach*, London: Longman.

MacLennan, D. (1986) *Maintenance and Modernisation of Urban Housing*, OECD Urban Affairs Programme.

de la Madrid, M. (1988) 'Sexto Informe de Gobierno', *Comercio Exterior*, 38, 771–86.

Malpezzi, S.J. (1986) 'Rent control and housing market equilibrium: theory and evidence from Cairo, Egypt', unpublished PhD thesis, George Washington University.

Malpezzi, S. and Mayo, S.K. (1987a) 'The demand for housing and developing countries: empirical estimates from household data', *Economic Development and Cultural Change*, 35, 687–721.

Malpezzi, S. and Mayo, S.K. (1987b) 'User cost and housing tenure in developing countries', *Journal of Development Economics*, 25, 197–220.

Malpezzi, S. and Rydell, C.P. (1986) *Rent Control in Developing Countries: a Framework for Analysis*, Washington: World Bank, Water Supply and Urban Development Department, Report No. UDD–102.

Mangin, W. (1967) 'Latin American squatter settlements: a problem and a solution', *Latin American Research Review*, 2, 65–98.

Margulis, M. (1981) 'Crecimiento y migración en una ciudad de la frontera: estudio preliminar de Reynosa', in R. González Salazar (ed.) *La Frontera del Norte: Integración y Desarrollo*, El Colegio de México.

Marroquín, E. (1985) 'Las vecindades de Puebla', in A. Gimate-Welsh and E. Marroquín *Lenguaje, Ideología y Clases Sociales. Las Vecindades en Puebla*, Editorial Universidad Autónoma de Puebla, 71–243.

Massolo, A. (1986) '"Que el gobierno entidenda, lo primero es la vivienda!"', *Revista Mexicana de Sociología*, 48, 195–238.

Mayo, S. (1985) 'How much will households spend for shelter?' *Urban Edge*, 9, 4–5.

Mele, P. (1984) *Los Procesos de Producción del Espacio Urbano en la Ciudad de Puebla*, Documento de Investigación No. 1, Instituto de Ciencias, Universidad Autónoma de Puebla.

Mele, P. (1985) *Cartografía Temática de la Ciudad de Puebla*, Universidad Autónoma de Puebla.

Mele, P. (1986a) *El Espacio Industrial entre la Ciudad y la Región*, Documento de Investigación No. 3, Instituto de Ciencias, Universidad Autónoma de Puebla.

Mele, P. (1986b) 'La dynamique de l'urbanisation de la ville de Puebla (Mexique) – de la ville a la région urbaine', unpublished PhD thesis, Université de Paris III.

Mele, P. (1987) 'Croissance urbaine, illegalité et pouvoir local dans la ville de Puebla (Mexique)', paper presented to Institute of British Geographers Conference, Portsmouth.

Mele, P. (1988a) 'Cartographier l'illegalité: filières de production de l'espace urbain de la ville de Puebla (Mexique)', *L'Espace Géographique*, 4, 257–63.

Mele, P. (1988b) 'Procesos de desarrollo espacial de la Ciudad de Puebla', paper presented at 46th Congress of Americanists, Amsterdam.

Méndez Rodríguez, A. (1987) 'Situación de la vivienda en arrendamiento y su impacto en el nivel de vida de los trabajadores', Instituto de Investigaciones Económicas, Universidad Nacional Autónoma de México, mimeo.

Méndez Sainz, E. (1987) *La Conformación de la Ciudad de Puebla (1966–80)*, Universidad Autónoma de Puebla.

Mexico, FONHAPO, (n.d.) *Políticas de administración creditica y financiera*.

Mexico, FOVI (Fondo de Operación y Financiamiento Bancario a la Vivienda) (1986) 'Programa Financiero de Vivienda', mimeo.

Mexico, FOVI (1989) *Sistema finaciero mexicano, 1982–88*.

Mexico, INC (Instituto Nacional del Consumidor) (1989) 'El gasto alimentario de la población de escasos recursos de la ciudad de México, *Comercio Exterior*, 39, 52–8.

Mexico, INEGI (Instituto Nacional de Estadística, Geografía e Informática) (1985) *Estadísticas Históricas de México Tomo 2*.

Mexico, INEGI (1989) *Agenda estadística 1988*.

Mexico, SAHOP (Secretaría de Asentamientos Humanos e Obras Públicas) (1978) *Desarrollo Urbano: Programa Nacional de Vivienda – Versión Abreviada*.

Mexico, SEDUE (Secretaría de Desarrollo Urbano e Ecología) (1987) *Vivienda: Decisiones Institucionales Programa 1987*.

Mexico, SEDUE (1989a) 'Estadística de vivienda 1983–88', mimeo.

Mexico, SEDUE (1989b) 'Foro de consulta popular sobre vivienda: conclusiones', mimeo.

Mexico, SHCP (Secretaría de Hacienda y Crédito Público) (1964) *Programa Financiero de Vivienda*.

Mexico, SPP (Secretaría de Programación y Presupuesto) (1982) *Trabajo y Salarios Industriales 1981*.

Michel, M.A. (1988) 'El proceso habitacional en la ciudad de México', in M.A. Michel, (ed.) *Procesos habitacionales de la ciudad de México*, Cuadernos Universitarios 51, SEDUE and UAM, 11–17.

Mohan, R. and Villamizar, R. (1982) 'The evolution of land values in the context of rapid urban growth: a case study of Bogotá and Cali, Colombia', in M. Cullen and S. Woolery (eds) *World Congress on Land Policy, 1980*, London: Lexington Press, 217–53.

Moitra, M.K. and Samajdar, S. (1987) 'Evaluation of the slum improvement programme of Calcutta bustees', in R. Skinner *et al.* (eds) *Shelter Upgrading for the Urban Poor: Evaluation of Third World Experience*, Manila: Island Publishing House, 69–86.

Morales, M.D. (1985) 'El proceso de demortización y de reforma y sus efectos en la distribución de la propiedad de la cuidad de México 1848–1864', paper presented to VII Reunión de Historiadores Mexicanos y Norteamericanos, Oaxaca.

Morfín, G.A. and Sánchez, M. (1984) 'Controles jurídicos y psicosociales en la producción de espacio urbano para sectores populares en Guadalajara', *Encuentro*, 1, 115–41.

Morgan, N.J. and Daunton, M.J. (1983) 'Landlords in Glasgow: a study of 1900', *Business History*, 25, 264–86.

Necochea, A. (1987) 'El allegamiento de los sin tierra, estrategia de supervivencia en vivienda', *Revista Latinoamericana de Estudios Urbano Regionales (EURE)*, 13–14, 85–100.

Negrete, M.A. and Salazar, H. (1986) 'Zonas metropolitanas en México, 1980', *Estudios Demográficos y Urbanos*, 1, 97–123.

Nelson, J.M. (1979) *Access to Power: Politics and the Urban Poor in Developing Nations*, Princeton, New Jersey: Princeton University Press.

Nelson, K.W. (1988) 'Choices and opportunities: low-income rental housing in Indonesia', *Urban Institute Working Paper*, 3780–812.

Nientied, P. and van der Linden, J.J. (1987) 'Evaluation of squatter settlement upgrading in Baldia, Karachi', in R. Skinner *et al.* (eds) *Shelter Upgrading for the Urban Poor: Evaluation of Third World Experience*, Manila: Island Publishing House, 107–126.

O'Connor, A.M. (1983) *The African City*, London: Hutchinson.

Okpala, D.C.I. (1985) 'Rent control reconsidered', in P. Onibokun (ed.) *Housing in Nigeria*, Ibadan: National Institute for Social and Economic Research, 139–59.

Ordoñez, J.L. (1989) *Vivir en Paz*, Cámara Nacional de Comercio de la Ciudad de México.

Ortiz Gil, C. (1982) *Viaje al Centro de un Submundo*, Mexico City: Castillo.

Parcero López, J. (1983) 'Iniciativa de Ley de regulación de rentas por parte del estado, sobre bienes inmuebles destinados para habitación, locales comerciales e industriales en pequeño', proposal prepared by the Diputados of the PRI of the 52nd Legislation, 1 August.

Payne, G. (1989) *Informal Housing and Land Subdivisions in Third World Cities: A Review of the Literature*, report prepared for the Overseas Development Administration, Oxford Polytechnic.

Peil, M. (1976) 'African squatter settlements: a comparative study', *Urban Studies*, 13, 155–66.

Peil, M. and Sada, P.O. (1984) *African Urban Society*, Chichester: John Wiley and Sons Ltd.

de la Peña, G. (1986) 'Mercados de trabajo y articulación regional: apuntes sobre el caso de Guadalajara y el occidente mexicano', in G. de la Peña and A. Escobar (eds), 47–88.

de la Peña, G. (1988) 'Movimientos sociales, intermediación política y poder local', paper presented to 46th International Congress of Americanists, Amsterdam.

de la Peña, G. and Escobar, A. (eds) (1986) *Cambio Regional, Mercado de Trabajo y Vida Obrera en Jalisco*, El Colegio de Jalisco.

Pérez Perdomo, R. and Nikken, P. (1982) 'The law and home ownership in the barrios of Caracas', in A.G. Gilbert, J.E. Hardoy, and R. Ramírez, (eds), 205–30.

Perló, M. (1979) 'Política y vivienda en México 1910–1952', *Revista Mexicana de Sociología*, 41, 769–835.

Perló, M. (1981) *Estado, Vivienda y Estructura Urbana en el Cardenismo: el Caso de la Ciudad de México*, Instituto de Investigaciones Sociales, Universidad Nacional Autónoma de México.

Portes, A. (1979) 'Housing policy, urban poverty, and the state: the favelas of Rio de Janeiro, 1972–76', *Latin American Research Review*, 14, 3–24.

Portillo, A.J. (1984) *El Arrendamiento de Vivienda en la Ciudad de México*, Cuadernos Universitarios No. 5, Universidad Autónoma Metropolitana-Iztapalapa.

Pozas Garza, M. (1989) 'Land settlement by the urban poor in Monterrey', in A.G. Gilbert (ed.), 65–78.

Preciado, E.S. and Ibañez, I. (1930) *Plano de la Ciudad de Guadalajara*.

Puebla, Gobierno del Estado and Gobierno del Municipio (1980) *Plan Director Urbano, Ciudad de Puebla*.

Puebla/SAHOP (1978) *Plan Estatal de Desarrollo Urbano: Puebla*, Gobierno del Estado de Puebla and SAHOP (Secretaría de Asentamientos Humanos e Obras Públicas).

Ramírez Jiménez, M. (1978) Vecindades en Guadalajara, Tesis profesional, Escuela de Arquitectura, ITESO.

Ramírez Saiz, J.M. (1986) 'Reivindiccaciones urbanas y organización popular. El caso de Durango', *Estudios Demográficos y Urbanos*, 1, 399–422.

Ray, T. (1969) *The Politics of the Barrios of Venezuela*, University of California Press.

Rébora Togno, A. (1986) 'Suelo urbano y desarrollo: el caso de México', paper given to Segundo Congreso Iberoamericano de Urbanismo, Tlaxcala, 21–25 April.

Regalado, J. (1987) 'El movimiento popular independiente en Guadalajara', in J. Tamayo (ed.) *Perspectivas de los Movimientos Sociales en la Región Centro-Occidente*, Mexico City: Editorial Linea, 121–57.

Renaud, B. (1987) 'Financing shelter', in L. Rodwin (ed.) *Shelter, Settlement and Development*, London: Allen & Unwin, 179–203.

Robben, P.J.M. (1987) 'Measurement of population dynamics following squatter settlement improvement in Ashok Nagar, Madras', in R. Skinner *et al.* (eds) *Shelter Upgrading for the Urban Poor: Evaluation of Third World Experience*, Manila: Island Publishing House, 87–106.

Rodas, F. and Sugranyes, A. (1988) 'El inquilinato en la ciudad de Guatemala: el caso de la zona 8', *Boletín de Medio Ambiente y Urbanización*, 7, 3–18.

Rodríguez Ortíz, J.L. (1984) 'Comentario' on E. Wario's 'Crecimiento urbano y acumulación de capital en el sector inmobiliario en el área urbana de Guadalajara', *Encuentro*, 1, 167–71.

Rosa, M. de la, (1985) *Marginalidad en Tijuana*, Tijuana: Centro de Estudios Fronterizos del Norte de México.

Sánchez, M. (1979) 'Le phenomène des fractionnements populaires a Guadalajara, Jalisco, Mexique', unpublished PhD thesis, Ecole des Hautes Etudes en Sciences Sociales, Paris.

Sanderson, S.R.W. (1984) *Land Reform in Mexico: 1910–1980*, London: Academic Press.

Sands, G. (1984) 'Into the nineties: review of Anthony Downs' *Rental Housing in the 1980s'*, *Growth and Change*, 15, 56.

Scobie, J. (1974) *Buenos Aires: Plaza to Suburb 1870–1910*, Oxford: Oxford University Press.

Scott, I. (1982) *Urban and Spatial Development in Mexico*, London: Johns Hopkins University Press.

Simpson, E.N. (1937) *The Ejido: Mexico's Way Out*, University of North Carolina Press.

Slater, D. (ed.) (1987) *New Social Movements and the State in Latin America*, Amsterdam: CEDLA publications.

Slovik, R. (1972) 'El inquilinato en el D.F.: un problema socio-económico', Tesis profesional, Escuela Nacional de Economía, Universidad Nacional Autónoma de México.

Stedman-Jones, G. (1971) *Outcast London*, Oxford University Press.

Sternlieb, G. (1969) *The Tenement Landlord*, New Brunswick, NJ: Rutgers University Press.

Sternlieb, G. and Hughes, J.W. (1981) 'The future of rental housing', Center for Urban Policy Research, Washington D.C, mimeo.

Stren, R.E. (1982) 'Underdevelopment, urban squatting, and the state bureaucracy: a case study of Tanzania', *Canadian Journal of African Studies*, 16, 67–91.

Sudra, T.L. (1976) 'Low-income housing system in Mexico City', unpublished doctoral thesis, Massachusetts Institute of Technology.

Sudra, T. (1984) 'Vivienda popular: mejoramiento progresivo o cinturón de miseria', *CEPES Jalisco*, 4, 67–75.

Sundaram, P.S.A. (1987) 'India', in S.-K. Ha (ed.) *Housing Policy and Practice in Asia*, London: Croom Helm.

Taibo, P.I. and Vizcaíno, R. (1984) 'Inquilinos del DF, a colgar la rojinegra', in P.I. Taibo and R. Vizcaíno (eds) *Memoria Roja: Luchas Sindicales de los años 20*, Mexico City: Ediciones Leega/Jucar, 147–83.

Tamayo, J.E. (1982) 'Apuntes para el estudio de la clase obrera y el movimiento sindical en Jalisco', *Relaciones: Estudios de Historia y Sociedad*, 3, 87–96.

Taschner, S.P. (1988) 'Diagnosis and challenges on housing in Brazil', paper presented to the Conference on Housing, Policy, Urban Innovation, Amsterdam, 27 June – 1 July.

Tipple, A.G. (1988) *The History and Practice of Rent Controls in Kumasi, Ghana*, Washington: World Bank, Water Supply and Urban Development Dept. Working Paper No. 88–1.

Trivelli, P. (1986) 'Access to land by the urban poor: an overview of the Latin American experience', *Land Use Policy*, 3, 101–21.

Turner, J.F.C. (1967) 'Barriers and channels for housing development in modernizing countries', *Journal of the American Institute of Planners*, 33, 167–81.

Turner, J.F.C. (1968) 'Housing priorities, settlement patterns, and urban development in modernizing countries', *Journal of the American Institute of Planners*, 34, 354–63.

Unikel, L., with Ruiz Chiappeto, C. and Garza, G. (1976) *El Desarrollo Urbano de México: Diagnóstico e Implicaciones Futuras*, El Colegio de México.

UN (United Nations) (1979) *Review of Rent Control in Developing Countries*, New York.

UN (United Nations) (1982) *United Nations Statistical Yearbook 1982*, New York.
UN (United Nations) (1985) *Compendium of Housing Settlement Statistics, 1982–84*, New York.
UNCHS (United Nations Centre for Human Settlements (HABITAT)) (1984) *Land for Human Settlements*, Nairobi: UNCHS.
UNCHS (United Nations, Centre for Human Settlements (HABITAT)) and IHS (Institute of Housing Studies) (1989) 'Conference on Rental Housing in Developing Countries', Rotterdam.
UNECLA (United Nations Economic Commission for Latin America) (1985) 'Preliminary overview of the Latin American Economy 1985', *Notas Sobre la Economía y el Desarrollo* 424/5.
UNECLA (United Nations Economic Commission for Latin America) (1986) *Statistical Yearbook for Latin America and the Caribbean 1985*, Santiago.
UNECLA (United Nations Economic Commission for Latin America) (1989) *Statistical Yearbook for Latin America and the Caribbean, 1988*, Santiago.
Universidad Autónoma de Puebla, Departamento de Investigaciones Arquitectónicas y Urbanísticas del Instituto de Ciencias (DIAU–ICUAP) (1984) *Crisis del Centro Histórico y las Vecindades*, Cuadernos de debate sobre problemas urbanos 4–5.
Universidad Autónoma de Puebla, Departamento de Investigaciones Arquitectónicas y Urbanísticas del Instituto de Ciencias (DIAU–ICUAP) (1986) *Memoria de la Primera Mesa de Trabajo sobre Investigaciones Universitarias de Urbanismo*.
Universidad de Guadalajara, Instituto de Asentamientos Humanos (1985) *Memoria Descriptiva – Bethel-Hernández Loza*, Estudios de Urbanismo Superior.
Urrutia, M. (1987) 'Latin America and the crisis of the 1980s', in L. Emmeris (ed.) *Development Policies and the Crisis of the 1980s*, Organisation of Economic Cooperation and Development, 56–69.
Varley, A. (1985a) 'Urbanisation and agrarian law: the case of Mexico City'. *Bulletin of Latin American Research*, 4, 1, 1–16.
Varley, A. (1985b) '*Ya somos dueños*: ejido land development and regularisation in Mexico City', unpublished PhD thesis, University College London.
Varley, A. (1989a) 'Propiedad de la Revolución? Los ejidos en el crecimiento de la ciudad de México', *Revista Interamericana de Planificación*, 22, 125–55.
Varley, A. (1989b) 'Settlement, illegality and legalization: the need for re-assessment', in P.M. Ward (ed.) *Corruption, Development and Inequality: Soft Touch or Hard Graft?*, London: Routledge, 156–74.
Vázquez, D. (1984) 'El sistema mixto para la captación de recursos y la toma de decisiones relativas al desarrollo urbano de Guadalajara. Un estudio de caso', *Encuentro*, 1, 87–108.
Vázquez, D. (1985) 'La ciudad en perspectiva', in P. Arias (ed.), 57–76.
Vázquez, D. (1989) 'Rural-urban land conversion on the periphery of Guadalajara', in A.G. Gilbert (ed.), 101–8.
Verbeek, H. (1987) 'The authorization of unauthorised housing in Cd. Chihuahua', in O. Verkoren and J. van Weesep (eds), 89–103.
Verkoren, O. and van Weesep, J. (eds) (1987) *Spatial Mobility and Urban Change*, Netherlands Geographical Studies 37.
Vernez, G. (1973) *The Residential Movements of Low-Income Families: the Case of Bogotá, Colombia*, Rand Institute.
Villareal, D.R. and Castañeda, V. (1986) *Urbanización y Autoconstrucción de Vivienda en Monterrey*, Mexico City: Centro de Ecodesarrollo.

Violich, F. (1944) *Cities of Latin America*, New York: Reinhardt, Holt & Winston.

Wahab, E.A. (1984) *The Tenant Market of Baldia Township: Towards a More General Understanding of Tenancy in Squatter Settlements*, Amsterdam: Free University Urban Research Working Paper No. 3.

Walton, J. (1977) *Elites and Economic Development: Comparative Studies on the Political Economy of Latin American Cities*, Institute of Latin American Studies, University of Texas at Austin.

Walton, J. (1978) 'Guadalajara: creating the divided city', *Latin American Urban Research*, 6, 25–50.

Ward, P.M. (1976) 'In search of a home: social and economic characteristics of squatter settlements and the role of self-help housing in Mexico City', unpublished PhD thesis, University of Liverpool.

Ward, P.M. (ed.) (1982) *Self-Help Housing: a Critique*, London: Mansell.

Ward, P.M. (1986) *Welfare Politics in Mexico*, London: Allen & Unwin.

Ward, P.M. (forthcoming) 'The politics of housing production in Mexico', in W. van Vliet (ed.) *The International Handbook of Housing Policies and Practices*, Connecticut: Greenwood Press.

Wario E. (1984) 'Crecimiento urbano y acumulación de capital en el sector inmobiliario en el área urbana de Guadalajara', *Encuentro*, 2, 146–66.

Wilkie, J.W. (1984) 'Changes in Mexico since 1895: central government revenue, public sector expenditure and national economic growth', in J.W. Wilkie and A. Perkal (eds) *Statistical Abstract of Latin America 23*, UCLA Latin American Center Publications, 861–80.

Winnie, W.W. (1987) *La Encuesta de Hogares de Guadalajara, 1986*, Universidad de Guadalajara, Instituto de Estudios Económicos y Regionales.

Wohl, A.S. (1971) 'The housing of the working classes in London, 1815–1914', in S.D. Chapman (ed.) *The History of Working-Class Housing: a Symposium*, Devon: David and Charles, 13–54.

World Bank (1980) *Shelter*, Poverty and Basic Needs Series, Washington DC.

World Bank (1982) *World Development Report 1982*, Oxford: Oxford University Press.

Yujnovsky, O. (1984) *Claves Políticas del Problema Habitacional Argentino*, Buenos Aires: Grupo Editor Latinoamericano.

Zavala, H.H. (1984) 'La oferta de suelo urbano y los sectores de bajos ingresos', *CEPES Jalisco*, 4, 76–9.

# Index

Printed and bound by CPI Group (UK) Ltd, Croydon, CR0 4YY

01/11/2024

01782616-0002